Hydrodynamics and Hydromagnetic Stability

Hydrodynamics and Hydromagnetic Stability

Gideon Paxton

RANDOM

RANDOM PUBLICATIONS

NEW DELHI (INDIA)

Hydrodynamics and Hydromagnetic Stability

ISBN 978-93-5111-957-9

Published in 2016 in India by

RANDOM PUBLICATIONS

4376-A/4B, Gali Murari Lal, Ansari Road
New Delhi-110 002
Phone : +9111-43580356, 011-23289044, 011-43142548
e-mail: sales@randompublications.com,
info@randompublications.com, randomexports@gmail.com

Type Setting by : Friends Media, Delhi-110089
Digitally Printed at : Replika Press Pvt. Ltd.

Preface

In fluid dynamics, hydrodynamic stability is the field which analyses the stability and the onset of instability of fluid flows. The study of hydrodynamic stability aims to find out if a given flow is stable or unstable, and if so, how these instabilities will cause the development of turbulence. The foundations of hydrodynamic stability, both theoretical and experimental, were laid most notably by Helmholtz, Kelvin, Rayleigh and Reynolds during the nineteenth century.

These foundations have given many useful tools to study hydrodynamic stability. These include Reynolds number, the Euler equations, and the Navier–Stokes equations. When studying flow stability it is useful to understand more simplistic systems, e.g. incompressible and inviscid fluids which can then be developed further onto more complex flows. Since the 1980s, more computational methods are being used to model and analyse the more complex flows.

In application to the magnetosphere boundary we find we can make some predictions without too precise a knowledge of all the parameters involved. At middle and low latitudes the first growing modes propagate across the Earth's field with a very low phase velocity and wave fronts closely aligned to meridian planes. The modes tend to exhibit circular polarisation in a plane almost perpendicular to the Earth's field. This behaviour should also occur at high latitudes when the magnetosheath field is closely aligned to he Earth's field.

The emphasis has been laid on fundamental ideas, so that the students may grasp the subject easily. Each topic has been introduced and developed in as simple and straightforward manner as possible to be well within the comprehension of students of all categories.

– Author

Contents

1

Fluid Flow and Measurement

INTRODUCTION

Unless otherwise noted, the types of fluid flows we will consider are those that are:

NON VISCOUS

There is no loss of energy due to internal friction between adjacent layers of the fluid.

STEADY STATE

At any point in the fluid, the velocity and pressure do not change with time — they have reached steady state equilibrium.

NON-TURBULENT

The fluid has no net angular momentum about any point in the fluid — there are no eddies in the flow. We usually call this type of flow to be a laminar flow.

These conditions are rarely encountered in the real world. For example, water flowing through a pipe will always have some drag force on it at the pipe boundary where the water is in contact with the wall of the pipe. None-the-less, the above condition can be and are a good first order approximation. A good course in Fluid Mechanics would not neglect the effects.

TYPES OF FLUID FLOW METERS

Differential Pressure Flowmeters

In a differential pressure drop device the flow is calculated by measuring the pressure drop over an obstructions inserted in the flow.

The differential pressure flowmeter is based on the Bernoullis Equation, where the pressure drop and the further measured signal is a function of the square flow speed.

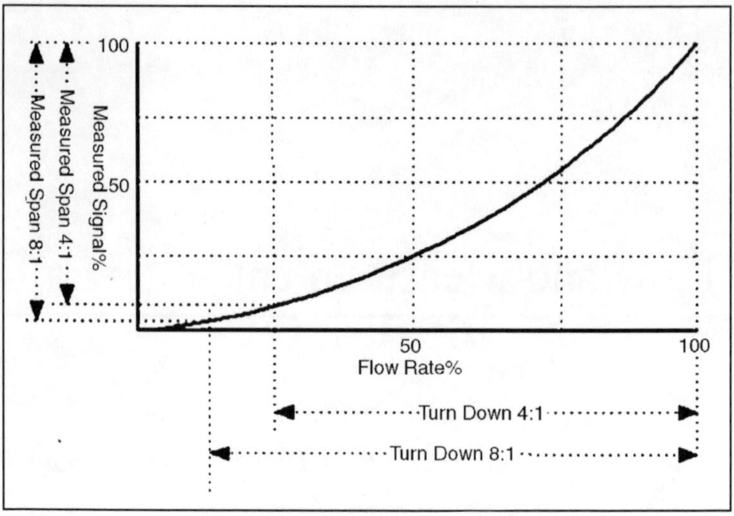

The most common types of differential pressure flowmeters are:

Orifice Plate

With an orifice plate, the fluid flow is measured through the difference in pressure from the upstream side to the downstream side of a partially obstructed pipe. The plate obstructing the flow offers a precisely measured obstruction that narrows the pipe and forces the flowing fluid to constrict.

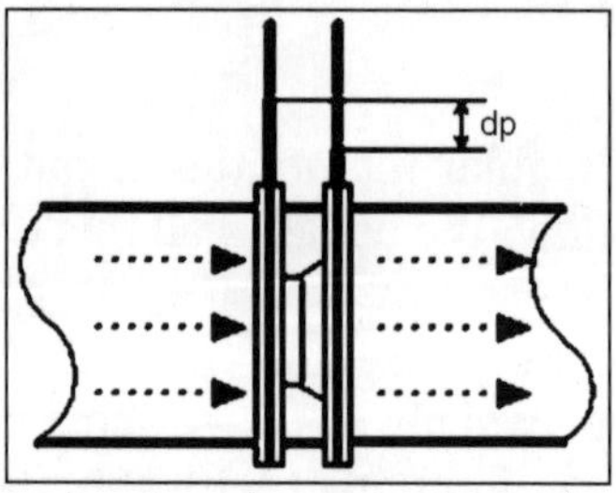

The orifice plates are simple, cheap and can be delivered for almost any application in any material.

The Turn Down Rate for orifice plates are less than 5:1. Their accuracy are poor at low flow rates. A high accuracy depend on an orifice plate in good shape, with a sharp edge to the upstream side. Wear reduces the accuracy.

- Orifice, Nozzle and Venturi Meters

VENTURI TUBE

Due to simplicity and dependability, the Venturi tube flowmeter is often used in applications where it's necessary with higher TurnDown Rates, or lower pressure drops, than the orifice plate can provide. In the Venturi Tube the fluid flowrate is measured by reducing the cross sectional flow area in the flow path, generating a pressure difference. After the constricted area, the fluid is passes through a pressure recovery exit section, where up to 80 per cent of the differential pressure generated at the constricted area, is recovered.

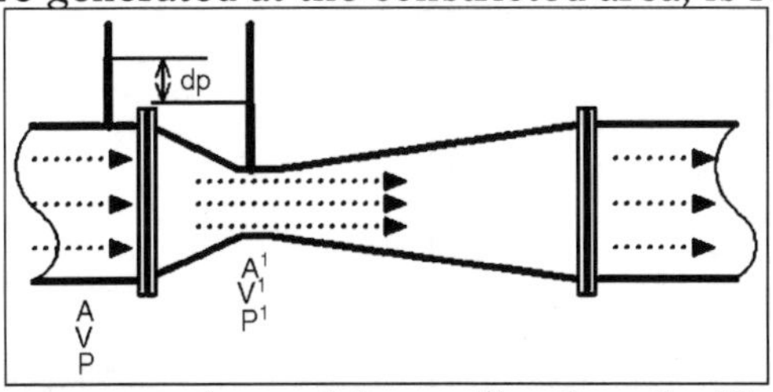

With proper instrumentation and flow calibrating, the Venturi Tube flowrate can be reduced to about 10 per cent of its full scale range with proper accuracy. This provides a TurnDown Rate 10:1.

Flow Nozzles

Flow nozzles are often used as measuring elements for air and gas flow in industrial applications.

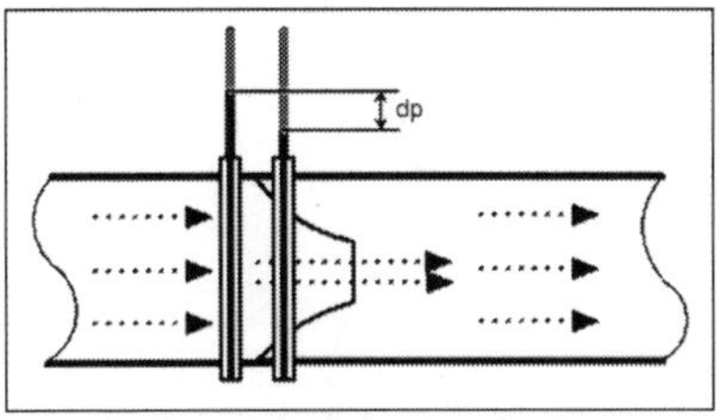

The flow nozzle is relative simple and cheap, and available for many applications in many materials.

The TurnDown Rate and accuracy can be compared with the orifice plate.

- Orifice, Nozzle and Venturi Meters

The Sonic Nozzle

When a gas accelerate through a nozzle, the velocity increase and the pressure and the gas density decrease. The maximum velocity is achieved at the throat, the minimum area, where it breaks Mach 1 or sonic. At this point it's not possible to increase the flow by lowering the downstream pressure. The flow is choked.

This situation is used in many control systems to maintain fixed, accurate, repeatable gas flow rates unaffected by the downstream pressure.

Nozzles and Venturi Meters

After the pressure difference has been generated in the differential pressure flow meter, the fluid pass through the pressure recovery exit section, where the differential pressure generated at the constricted area is partly recovered.

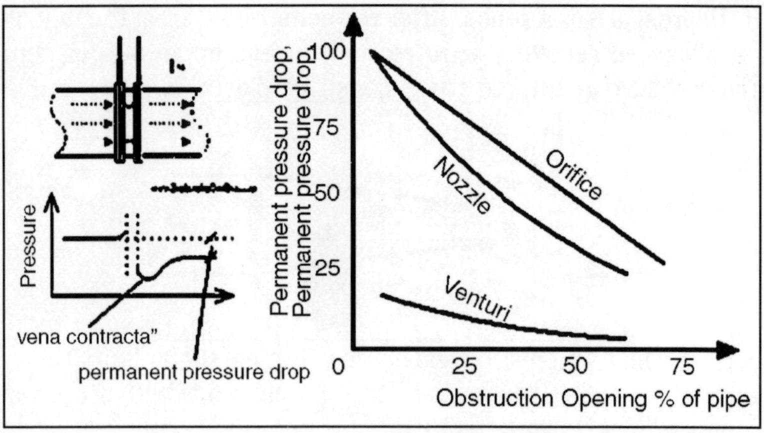

As we can see, the pressure drop in orifice plates are significant higher than in the venturi tubes.

VARIABLE AREA FLOWMETER OR ROTAMETER

The rotameter consists of a vertically oriented glass (or plastic) tube with a larger end at the top, and a metering float which is free to move within the tube. Fluid flow causes the float to rise in the tube as the upward pressure differential and buoyancy of the fluid overcome the effect of gravity.

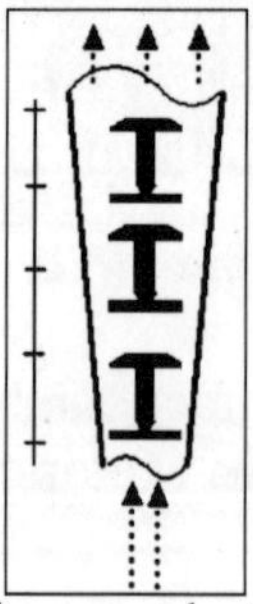

The float rises until the annular area between the float and tube increases sufficiently to allow a state of dynamic equilibrium between the upward differential pressure and buoyancy factors, and downward gravity factors.

The height of the float is an indication of the flow rate. The tube can be calibrated and graduated in appropriate flow units. The rotameter meter typically have a TurnDown Ratio up to 12:1. The accuracy may be as good as 1 per cent of full scale rating. Magnetic floats can be used for alarm and signal transmission functions.

VELOCITY FLOWMETERS

In a velocity flowmeter the flow is calculated by measuring the speed in one or more points in the flow, and integrating the flow speed over the flow area.

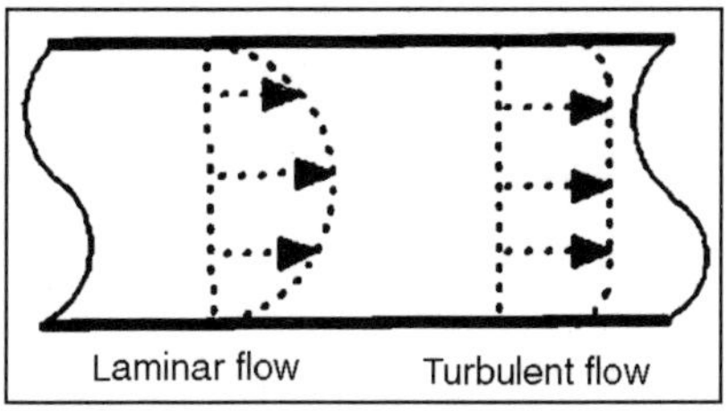

Pitot Tubes

The pitot tube are one the most used (and cheapest) ways to measure fluid flow, especially in air applications as ventilation and HVAC systems, even used in airplanes for the speed measurent.

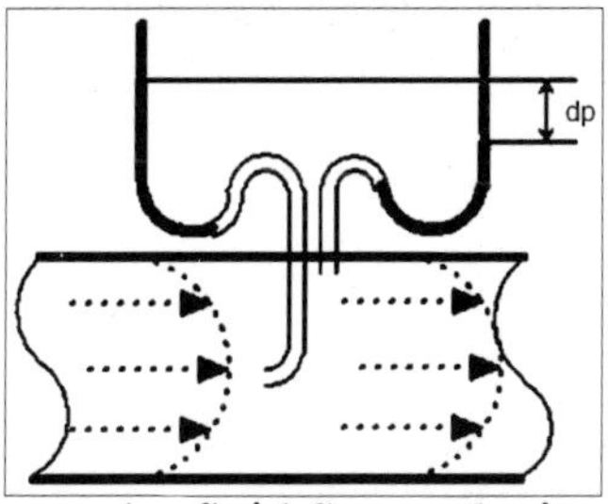

The pitot tube measures the fluid flow velocity by converting the kinetic energy of the flow into potential energy.

The use of the pitot tube is restricted to point measuring. With the "annubar", or multi-orifice pitot probe, the dynamic pressure can be measured across the velocity profile, and the annubar obtains an averaging effect.

Calorimetric Flowmeter

The calorimetric principle for fluid flow measurement is based on two temperature sensors in close contact with the fluid but thermal insulated from each other.

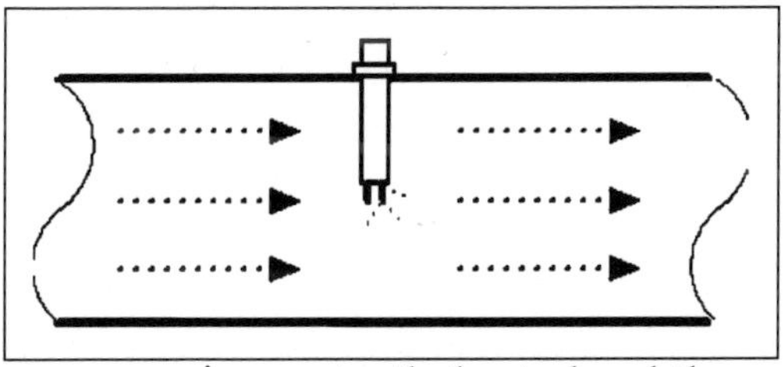

One of the two sensors is constantly heated and the cooling effect of the flowing fluid is used to monitor the flowrate. In a stationary (no flow) fluid condition there is a constant temperature difference between the two temperature sensors. When the fluid flow increases, heat energy is drawn from the heated sensor and the temperature difference between the sensors are reduced. The reduction is proportional to the flow rate of the fluid.

Response times will vary due the thermal conductivity of the fluid. In general lower thermal conductivity require higher velocity for proper measurement. The calorimetric flowmeter can achieve relatively high accuracy at low flow rates.

Turbine Flowmeter

There is many different manufacturing design of turbine flow meters, but in general they are all based on the same simple principle: If a fluid moves through a pipe and acts on the vanes of a turbine, the turbine will start to spin and rotate. The rate of spin is measured to calculate the flow.

The turndown ratios may be more than 100:1 if the turbine meter is calibrated for a single fluid and used at constant conditions. Accuracy may be better than +/-0,1 per cent.

Vortex Flow Meter

An obstruction in a fluid flow creates vortices in a downstream flow. Every obstruction has a critical fluid flow speed at which vortex shedding occurs. Vortex shedding is the instance where alternating low pressure zones are generated in the downstream.

These alternating low pressure zones cause the obstruction to move towards the low pressure zone. With sensors gauging the vortices the strength of the flow can be measured.

- The Vortex Flowmeter Principle - An introduction to the vortex flowmeter principle.

Electromagnetic Flowmeter

An electromagnetic flowmeter operate on Faraday's law of electromagnetic induction that states that a voltage will be induced when a conductor moves through a magnetic field. The liquid serves as the conductor and the magnetic field is created by energized coils outside the flow tube.

The voltage produced is directly proportional to the flow rate. Two electrodes mounted in the pipe wall detect the voltage which is measured by a secondary element.

Electromagnetic flowmeters can measure difficult and corrosive liquids and slurries, and they can measure flow in both directions with equal accuracy.

Electromagnetic flowmeters have a relatively high power consumption and can only be used for electrical conductive fluids as water.

- The Electromagnetic Flowmeter Principle - An introduction to the electromagnetic flowmeter principle

Ultrasonic Doppler Flowmeter

The effect of motion of a sound source and its effect on the frequency of the sound was observed and described by Christian Johann Doppler.

The frequency of the reflected signal is modified by the velocity and direction of the fluid flow

If a fluid is moving towards a transducer, the frequency of the returning signal will increase. As fluid moves away from a transducer, the frequency of the returning signal decrease.

The frequency difference is equal to the reflected frequency minus the originating frequency and can be use to calculate the fluid flow speed.

- The Ultrasonic Doppler and Time of Flight Flowmeter
- An Ultrasonic Flowmeter Tutorial - A basic tutorial about ultrasonic flowmeters.

Positive Displacement Flowmeter

The positive displacement flowmeter measures process fluid flow by precision-fitted rotors as flow measuring elements. Known and fixed volumes are displaced between the rotors.

The rotation of the rotors are proportional to the volume of the fluid being displaced.

The number of rotations of the rotor is counted by an integral electronic pulse transmitter and converted to volume and flow rate.

The positive displacement rotor construction can be done in several ways:

- *Reciprocating piston* meters are of single and multiple-piston types.
- Oval-gear meters have two rotating, oval-shaped gears with synchronized, close fitting teeth. A fixed quantity of liquid passes through the meter for each revolution. Shaft rotation can be monitored to obtain specific flow rates.
- *Nutating disk* meters have moveable disks mounted on a concentric sphere located in spherical side-walled chambers. The pressure of the liquid passing through the measuring chamber causes the disk to rock in a circulating path without rotating about its own axis. It is the only moving part in the measuring chamber.
- *Rotary vane* meters consists of equally divided, rotating impellers, two or more compartments, inside the meter's housings. The impellers are in continuous contact with the casing. A fixed volume of liquid is swept to the meter's outlet from each compartment as the impeller rotates. The revolutions of the impeller are counted and registered in volumetric units.

The positive displacement flowmeter may be used for all relatively non-abrasive fluids such as heating oils, lubrication oils, polymer additives, animal and vegetable fat, printing ink, freon, and many more. Accuracy may be up to *0.1* per cent of full rate with a TurnDown of 70:1 or more.

Mass Flowmeters

Mass meters measure the mass flow rate directly.

Thermal Flowmeter

The thermal mass flowmeter operates independent of density, pressure, and viscosity. Thermal meters use a heated sensing element isolated from the fluid flow path where the flow stream conducts heat from the sensing element. The conducted heat is directly proportional to the mass flow rate and the he temperature difference is calculated to mass flow.

The accuracy of the thermal mass flow device depends on the calibrations reliability of the actual process and variations in the temperature, pressure, flow rate, heat capacity and viscosity of the fluid.

Coriolis Flowmeter

Direct mass measurement sets Coriolis flowmeters apart from other technologies. Mass measurement is not sensitive to changes in pressure, temperature, viscosity and density. With the ability to measure liquids, slurries and gases, Coriolis flowmeters are universal meters.

Flowmeter uses the Coriolis effect to measure the amount of mass moving through the element. The fluid to be measured runs through a U-shaped tube that is caused to vibrate in an angular harmonic oscillation.

Due to the Coriolis forces, the tubes will deform and an additional vibration component will be added to the oscillation.

This additional component causes a phase shift on some places of the tubes which can be measured with sensors. The Coriolis flow meters are in general very accurate, better than +/-0,1 per cent with an turndown rate more than 100:1. The Coriolis meter can also be used to measure the fluids density.

Open Channel Flowmeters

A common method of measuring flow through an open channel is to measure the height of the liquid as it passes over an obstruction as a flume or weir in the channel.

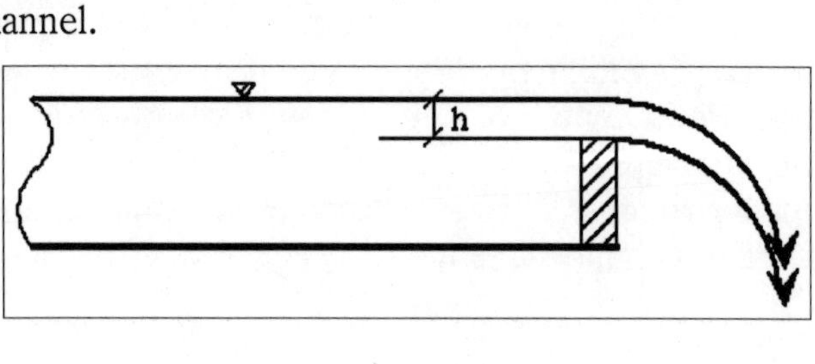

Common used is the Sharp-Crested Weir, the V-Notch Weir, the Cipolletti weir, the Rectangular-Notch Weir, the Parshall Flume or Venturi Flume.

TYPES OF FLUID MOTION

Fluids, like most other forms of matter, are made up of tiny particles [molecules], which are separated by large spaces.

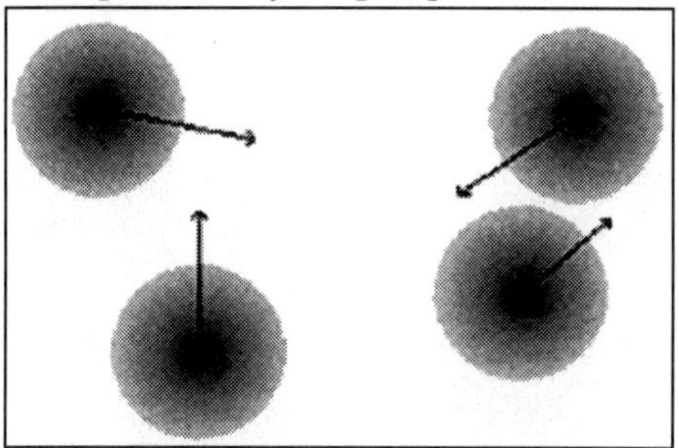

On a really tiny level, we would have to deal with these individual particles. Fortunately, most of the problems we encounter in aerodynamics can be handled by looking at volumes no smaller than, say, (1×10^{-6} m^3), a volume which contains at least a few thousand molecules. [How to calculate: From Chemistry: Avogadro's # gives # of atoms (or molecules) per gram-mole of the gas.

Density of air (mass/volume) can be found from the Perfect Gas Law:

$$P = rRT$$

where R is gas constant (~ 288 in SI units), and T is temperature, and P is pressure.

⇒ *So we can treat fluid as a "continuum"*: A medium which is pretty uniform in properties at the smallest scales of interest to us.

When one tries to analyse the dynamics of gases at the outer edge of earth's atmosphere, one has to look for methods which don't assume "continuum" fluid mechanics: the distances between molecules are of the same order as the dimensions of interest to us; perhaps meters.

The same problem arises at the other end of the size scale: when we try to deal with micro-devices used to sense and control flow behaviour. Here, even at usual sea-level atmospheric conditions, the dimensions of the device may be so small that that they are comparable to the distance between molecules.

Fluid Motion

In the rest of this course we will treat fluids as "continuous media", and forget about those molecules zipping about at random. The smallest "packet" of fluid which we will now consider will have many millions of molecules. The motion of these packets is the net motion of all those molecules zipping around within the packet (think of a school bus moving at 5mph with thirty middle-schoolers conducting a football game inside using each other's backpacks.)

Streamline

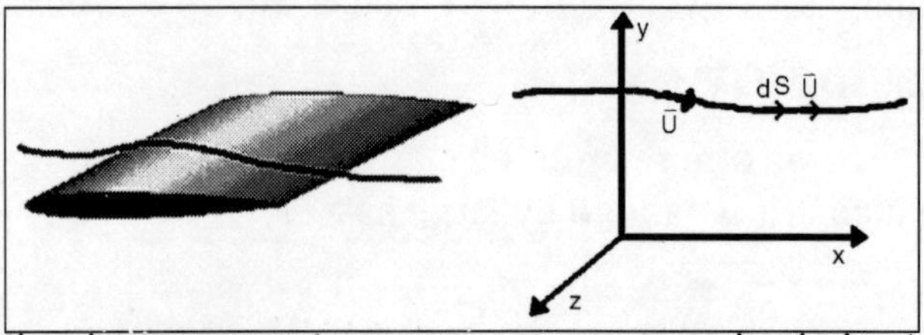

A streamline is a curve whose tangent at any point is in the direction of the velocity vector at that point.

We se that if the flow is not steady, the streamline has little use or significance.

If ds is an elemental vector along the streamline,

$$d\vec{s}=dx\,\vec{i}+dy\,\vec{j}+dz\,\vec{k}$$

$$\vec{U}=u\,\vec{i}+v\,\vec{j}+w\,\vec{k}$$

$$d\vec{s}\times\vec{U}=\begin{vmatrix} i & j & k \\ dx & dy & dz \\ u & v & w \end{vmatrix}$$

$$\Rightarrow i(wdy-vdz)+j(udz-wdx)+k(vdx-udy)=0$$

$$wdy=vdz\,\frac{dz}{dy}=\frac{w}{v}$$

$$udz=wdx\,\frac{dz}{dx}=\frac{w}{u}$$

Equation to a streamline is:

$$vdx=udy\,\frac{dy}{dx}=\frac{v}{u}$$

As a "packet" of fluid moves along, a combination of 4 things can happen to it: translation, dilatation, rotation and shear.

FLUID MOTION

Thus Basic fluid motion can be described as some combination of

Translation

This is characterized by the velocity

$$\vec{U}=u\,\vec{i}+v\,\vec{j}+w\,\vec{k}$$

$$\rho=\frac{P}{RT}$$

Dilatation

[volume change]

$$\vec{u}, \rho, p, e, h$$

Rotation

[About one, two or 3 axes].

$$\omega = \frac{1}{2}\left[\left(\frac{dw}{dy} - \frac{dv}{dz}\right)\vec{i} + \left(\frac{du}{dz} - \frac{dw}{dx}\right)j + \left(\frac{dv}{dx} - \frac{du}{dy}\right)k\right]$$

Note that rotation is a vector. We usually use a quantity called "vorticity" which is twice the rotation vector, to describe the amount of rotation in flows.

Shear Strain

The quantity "strain" thus has nine components.

They are:

- e_{xx}, e_{yy}, e_{zz} These are "normal strain" components.
- e_{xy}, e_{yz}, e_{zx} These are "shear strain" components.
- e_{yx}, e_{zy}, e_{xz} These are equal to the corresponding "shear strain" components above.

If this discussion brings back memories of your solid mechanics / statics course, there is good reason for it. The difference comes in the relation between strain and the stress: in the case of solids, the stress depends on the strain. In the case of fluids, the stress depends on the rate of strain.

Circulation

Defined as $d\vec{s}$ integrated around a closed contour. The negative sign is included such that positive circulation on a body corresponds to positive lift, and the integral is evaluated counter-clockwise.

From the preceding discussion, we see that r will be zero unless there is some vorticity, $\vec{\omega}$ contained within the contour. *r is an extremely useful quantity*: it helps us calculate lift, vortex strength, etc.

Important Points

$$d\vec{s} \times \vec{U} = 0$$

The circulation around a closed contour with net rotation and/or shear will be non-zero. It is, however, always possible to have a combination of rotation and/or shear that gives a zerocirculation.

CLASSIFICATION OF FLUID FLOW

Fluid flow can be very smooth, calm, and regular, but generally the flow of a fluid is not so disciplined- it becomes a vagabond and starts flowing in random patterns. What are different flow patterns, how are flows classified, and what are the fluid paths in those patterns? While studying the motion of a rigid body we do not have to bother about the relative motion of the particles of the rigid body as they are very firmly fixed to each other and move as a whole. But for the study of the motion of fluids, things are not so simple because the fluid particles are attached with each other with very weak forces. There are various relative motions and a lot of possibilities for relative motion between the fluid particles.

To make things somewhat simple or for making the flow analysis feasible, fluid flow is visualized as a composition of fluid elements. These elements are defined by using certain similarities or patterns and mathematics is applied to them to study fluid flow comprehensively.

Rotational or Irrotational Flow

To classify any flow as Rotational or Irrotational the angular motion of the fluid elements is analysed. If the angle between the two intersecting lines of the boundary of the fluid element changes while moving in the flow, then the flow is a Rotational Flow. But if the fluid element rotates as a whole and there is no change in angles between the boundary lines then the flow cannot be Rotational Flow, so it is Irrotational Flow.

This means that there should be some deformation in the fluid element in a Rotational Flow. Such deformation of the fluid element or the shear strain is necessarily caused by tangential forces or shear stresses. Shear stresses are caused by viscosity, thus the flow of viscous fluids is rotational. But this does not mean that the flow of non-viscous or ideal fluid is always irrotational. The flow of ideal fluids can be rotational by external work or heat interaction.

Laminar or Turbulent Flow

Laminar Flow

The flow of a fluid moving with a moderate speed has fluid layers moving past other layers as if some sheets are moving over other layers. Such flow of fluids is called Laminar Flow. In Laminar Flow viscous shear stresses act between these layers of the fluid which defines the velociity distribution among these layers of flow. In Laminar Flows the shear stresses are defined by Newton's equation for shear stress.

Turbulent Flow

As the flow speed of the otherwise calm layers increases, these smoothly moving layers start moving randomly, and with further increase in flow velocity, the flow of fluid particles becomes complctely random and no such laminar layers exist any more. Shear stresses in the Turbulent Flow are more than those in Laminar Flow.

A dimensionless parameter, Reynolds Number, is defined as the ratio of inertial and viscous force to characterize these two types of flow patterns. With increase in flow velocity the intial forces increase so the Reynonlds Number.

For moderate flows the Reynolds Number is below 2000 and for Turbulent Flows it is well above 2300. For the transition region between the two types the Reynolds Number varies between 2000-4000.

Fundamentals of Fluid Flow

How does a flowing fluid behaves,and how do the different parameters vary in a flowing fluid? Fluid flow analysis involves theoretical and experimental study of the flow of fluid by defining flow models and observation of variation of parameters. This series will discuss Fundamentals of Fluid Flow.

CONCEPT OF FLUID

What is fluid mechanics? As its name suggests it is the branch of applied mechanics concerned with the statics and dynamics of fluids - both liquids and gases. The analysis of the behaviour of fluids is based on the fundamental laws of mechanics which relate continuity of mass and energy with force and momentum together with the familiar solid mechanics properties.

We normally recognise three states of matter: solid; liquid and gas. However, liquid and gas are both fluids: in contrast to solids they lack the ability to resist deformation. Because a fluid cannot resist the deformation force, it moves, it *flows* under the action of the force. Its shape will change continuously as long as the force is applied. A solid can resist a deformation force while at rest, this force may cause some displacement but the solid does not continue to move indefinitely.

The deformation is caused by *shearing* forces which act tangentially to a surface. Referring to the figure below, we see the force F acting tangentially on a rectangular (solid lined) element ABDC. This is a shearing force and produces the (dashed lined) rhombus element A'B'DC.

$$\sigma = \frac{\rho \text{subtance}}{\rho \text{H}_2\text{O(at 4°C)}}$$

Shearing force, F, acting on a fluid element.

We can then say:

A Fluid is a substance which deforms continuously, or flows, when subjected to shearing forces.

and conversely this definition implies the very important point that:

If a fluid is at rest there are no shearing forces acting.

All forces must be perpendicular to the planes which the are acting.

When a fluid is in motion shear stresses are developed if the particles of the fluid move relative to one another. When this happens adjacent particles have different velocities. If fluid velocity is the same at every point then there is no shear stress produced: the particles have zero *relative* velocity.

Consider the flow in a pipe in which water is flowing. At the pipe wall the velocity of the water will be zero. The velocity will increase as we move towards the centre of the pipe. This change in velocity across the direction of flow is known as velocity profile and shown graphically in the figure below:

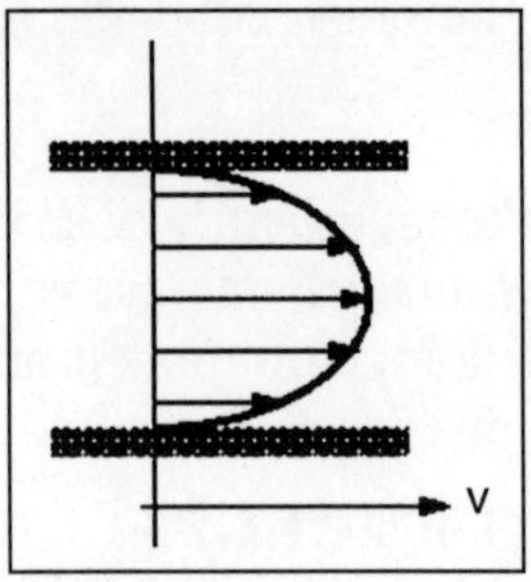

Fig. Velocity Profile in a Pipe.

Because particles of fluid next to each other are moving with different velocities there are shear forces in the moving fluid *i.e.* shear forces are normally present in a moving fluid. On the other hand, if a fluid is a long way from the boundary and all the particles are travelling with the same velocity, the velocity profile would look something like this:

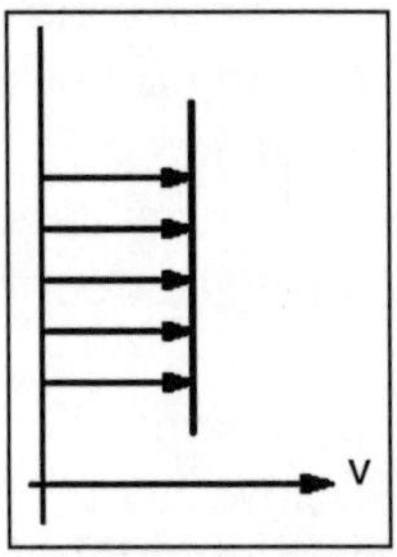

Fig. Velocity Profile in Uniform Flow

and there will be no shear forces present as all particles have zero relative velocity. In practice we are concerned with flow past solid boundaries; aeroplanes, cars, pipe walls, river channels etc. and shear forces will be present.

1. *Newton's law of viscosity*: How can we make use of these observations? We can start by considering a 3d rectangular element of fluid, like that in the figure.

   The shearing force F acts on the area on the top of the element. This area is given by $A = \delta s \times \delta x$. We can thus calculate the *shear stress* which is equal to force per unit area *i.e.*

$$\text{shear stress}, \tau = \frac{F}{A}$$

   The deformation which this shear stress causes is measured by the size of the angle f and is know as *shear strain*.

   In a solid shear strain, f, is constant for a fixed shear stress t.

In a fluid f increases for as long as t is applied - the fluid flows.

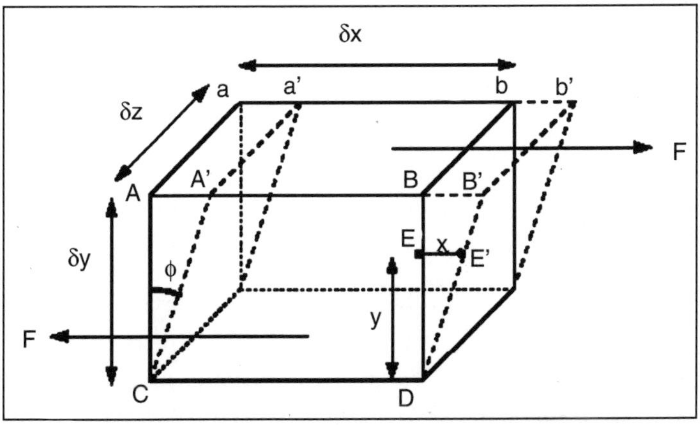

Fig. Fluid Element Under a Shear Force

It has been found experimentally that the *rate of shear stress* (shear stress per unit time, t/time) is directly proportional to the shear stress.

If the particle at point E (in the above figure) moves under the shear stress to point E' and it takes time *t* to get there, it has moved the distance *x*. For small deformations we can write shear strain,

$$\phi = \frac{x}{y}$$

$$\text{rate of shear strain} = \frac{\phi}{t}$$

$$= \frac{x}{ty} = \frac{x}{t}\frac{1}{y}$$

$$= \frac{u}{y}$$

where,

$$\frac{x}{t} = u$$

is the velocity of the particle at E.

Using the experimental result that shear stress is proportional to rate of shear strain then,

$$\tau = \text{constant} \times \frac{u}{y}$$

The term,

$$\frac{u}{y}$$

is the change in velocity with y, or the velocity gradient, and may be written in the differential form,

$$\frac{du}{dy}$$

The constant of proportionality is known as the dynamic viscosity, μ, of the fluid, giving

$$\tau = \mu \frac{du}{dy}$$

This is known as Newton's law of viscosity.

2. *Fluids vs. Solids*: Earlier we have discussed the differences between the behaviour of solids and fluids under an applied force. Summarising, we have;
   - For a solid the strain is a function of the applied stress (providing that the elastic limit has not been reached). For a fluid, the rate of strain is proportional to the applied stress.
   - The strain in a solid is independent of the time over which the force is applied and (if the elastic limit is not reached) the deformation disappears when the force is removed. A fluid continues to flow for as long as the force is applied and will not recover its original form when the force is removed.

   It is usually quite simple to classify substances as either solid or liquid. Some substances, however, (*e.g.* pitch or glass) appear solid under their own weight. Pitch will, although appearing solid at room temperature, deform and spread out over days - rather than the fraction of a second it would take water.

   Looking at properties of solids, when the elastic limit is reached they seem to flow. They become plastic. They still do not meet the definition of true fluids as they will only flow after a certain minimum shear stress is attained.

3. *Newtonian / Non-Newtonian Fluids*: Even among fluids which are accepted as fluids there can be wide differences in behaviour under stress. Fluids obeying Newton's law where the value of m is constant are known as Newtonian fluids. If m is constant the shear stress is linearly dependent on velocity gradient. This is true for most common fluids.

   Fluids in which the value of m is not constant are known as non-Newtonian fluids. There are several categories of these, and they are outlined briefly below.

These categories are based on the relationship between shear stress and the velocity gradient (rate of shear strain) in the fluid. These relationships can be seen in the graph below for several categories

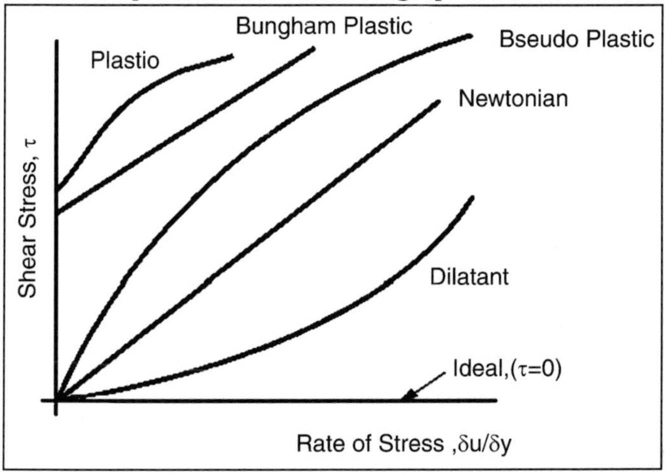

Fig. Shear Stress vs. Rate of Shear Strain du/dy

Each of these lines can be represented by the equation,

$$\tau = A + B\left(\frac{\delta u}{\delta y}\right)^n$$

where A, B and n are constants. For Newtonian fluids A = 0, B = m and n = 1.

Below are brief description of the physical properties of the several categories:

- *Plastic:* Shear stress must reach a certain minimum before flow commences.
- *Bingham plastic:* As with the plastic above a minimum shear stress must be achieved. With this classification n = 1. An example is sewage sludge.
- *Pseudo-plastic:* No minimum shear stress necessary and the viscosity decreases with rate of shear, *e.g.* colloidial substances like clay, milk and cement.
- *Dilatant substances;* Viscosity increases with rate of shear *e.g.* quicksand.
- *Thixotropic substances:* Viscosity decreases with length of time shear force is applied *e.g.* thixotropic jelly paints.
- *Rheopectic substances:* Viscosity increases with length of time shear force is applied
- *Viscoelastic materials:* Similar to Newtonian but if there is a sudden large change in shear they behave like plastic.

There is also one more - which is not real, it does not exist - known as the ideal fluid. This is a fluid which is assumed to have no viscosity. This is a useful concept when theoretical solutions are being considered - it does help achieve some practically useful solutions.

4. *Liquids vs. Gasses*: Although liquids and gasses behave in much the same way and share many similar characteristics, they also possess distinct characteristics of their own. Specifically,
   - A liquid is difficult to compress and often regarded as being incompressible. A gas is easily to compress and usually treated as such - it changes volume with pressure.
   - A given mass of liquid occupies a given volume and will occupy the container it is in and form a free surface (if the container is of a larger volume).

   A gas has no fixed volume, it changes volume to expand to fill the containing vessel. It will completely fill the vessel so no free surface is formed.
5. *Causes of Viscosity in Fluids*:
   - *Viscosity in Gasses*: The molecules of gasses are only weakly kept in position by molecular cohesion (as they are so far apart). As adjacent layers move by each other there is a continuous exchange of molecules. Molecules of a slower layer move to faster layers causing a drag, while molecules moving the other way exert an acceleration force. Mathematical considerations of this momentum exchange can lead to Newton law of viscosity.

   If temperature of a gas increases the momentum exchange between layers will increase thus increasing viscosity.

   Viscosity will also change with pressure - but under normal conditions this change is negligible in gasses.
   - *Viscosity in Liquids*: There is some molecular interchange between adjacent layers in liquids - but as the molecules are so much closer than in gasses the cohesive forces hold the molecules in place much more rigidly. This cohesion plays an important roll in the viscosity of liquids.

   Increasing the temperature of a fluid reduces the cohesive forces and increases the molecular interchange. Reducing cohesive forces reduces shear stress, while increasing molecular interchange increases shear stress. Because of this complex interrelation the effect of temperature on viscosity has something of the form:

$$\mu_T = \mu_0 \left(1 + AT + BT\right)$$

   where μ_T is the viscosity at temperature TC, and μ_0 is the viscosity at temperature 0C. A and B are constants for a particular fluid.

High pressure can also change the viscosity of a liquid. As pressure increases the relative movement of molecules requires more energy hence viscosity increases.

MEASURE OF RESISTANCE TO FLOW

VISCOSITY

The *viscosity* is a measure of resistance to flow. It we tip over a glass of water on the table, the water will spill out before we can stop it.

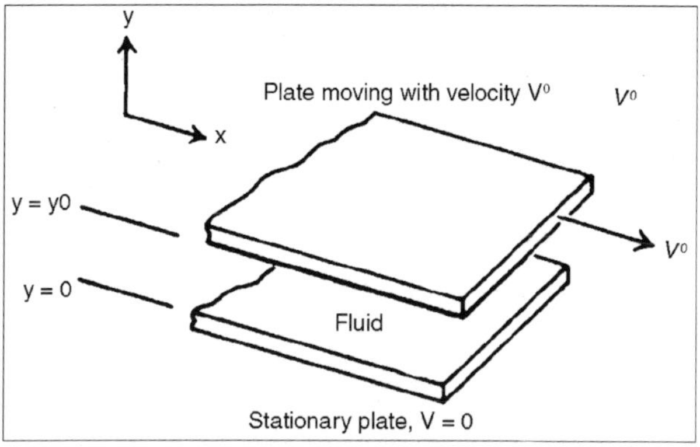

Fig. The Sliding Plate Experiment.

If we tip over a jar of honey, we probably can set it upright again before much honey flow out; this is because the honey has much greater resistance to flow, more viscosity, than water. A more precise definition of viscosity is possible in terms of following experiment:

Consider two long, solid plates separated by a thin film of fluid. If we slide the upper plate steadily in the *x* direction with velocity V_0, a force is needed to overcome the friction in the fluid between the plates. This force is different for different velocities, plate sizes, fluids, and distances between the plates. We can eliminate the effect of different plate sizes, however, by measuring the force per unit area of the plate, which we define as the shear stress τ.

It has been demonstrated experimentally that at low values of V_0 the velocity profile in the fluid between the plates is linear, *i.e.*,

$$V = \frac{V_o y}{y_o}$$

so that,

$$\frac{dV}{dy} = \frac{V_o}{y_o}$$

It has also been demonstrated experimentally that for most fluids the results of this experiment can be shown most conveniently on a plot τ versus dV/ dy. As shown here, dV/ dy is simply is velocity divided by a distance. In more complex geometries, it is the limiting value of such a ratio at a point. It is commonly called the *shear rate*, *rate of strain*, and *rate of shear deformation*, which all means exactly the same thing. Four different kinds of curve are shown as experiment results in the figure.

All four are observed in nature. The behaviour most common in nature is that represented by the straight line through the origin. This line is called *Newtonian* because it is described by Newton's law of viscosity,

$$\iota = \upsilon \frac{dV}{dy}$$

This equation says that the shear stress τ is linearly proportional to the velocity gradient dV/ dy. It is also the definition of viscosity, because we can rearrange it as,

$$\upsilon \frac{\iota}{dV/dy}$$

Here μ is called the *viscosity* or the coefficient of *viscosity*.

For fluids such as air, the value of μ is very low; therefore, their observed behaviour is represented in Fig above by a straight line through the origin, very close to the dV/dy axis. For fluid such as corn syrup, the value of μ is very large, and the straight line through the origin is close to the τ axis.

Fluids that exhibit this behaviour in the sliding plate experiment (*i.e.*, fluids that obey Newton's law of viscosity) are called *Newtonian fluids*.

All others are called *non-newtonian fluids*. Which fluids are Newtonian? All gases are Newtonian. All liquids for which we can write a simple chemical formula are Newtonian, such as water, benzene, ethyl alcohol, carbon tetrachloride, and hexane. Most solutions of simple molecules are Newtonian, such as aqueous solutions of inorganic salts and of sugar.

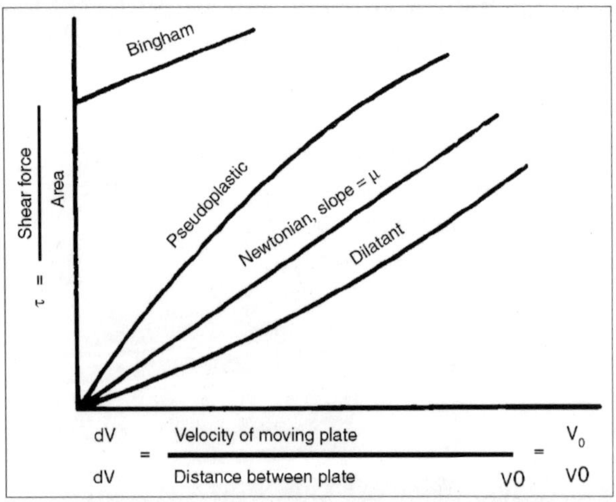

Fig. Possible outcomes of the Sliding-plate Experiment at Constant Temperature and Pressure.

Which fluids are non-newtonian? Generally, non-newtonian fluids are complex mixtures: slurries, pastes, gels, polymer solutions, etc. Most non-newtonian fluids are composed of molecules or particles that are much larger than water molecules, such as the sand grains in a mud or the collagen molecules in gelatin, which are thousands or millions of times larger than water molecules.

In discussing non-newtonian fluids, we must agree on what we mean by viscosity. If we retain the definition given by Eqn. then the viscosity can no longer be considered a constant independent of *dV/dy* for a given temperature, but must be considered a function of *dV/dy*. This is shown in figure below.

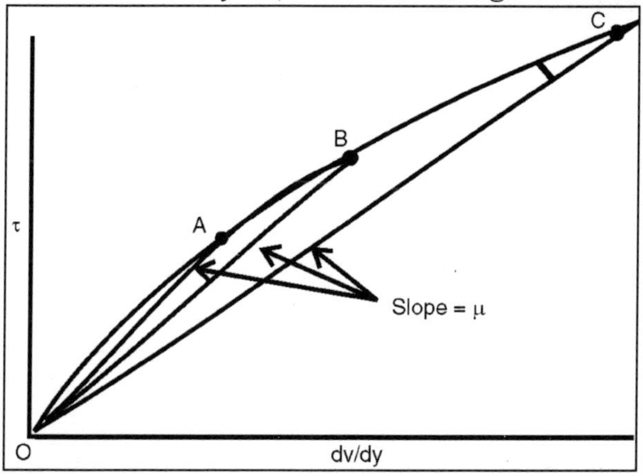

Fig. Viscosity of Non-newtonian Fluids at Constant Temperature

Here each of lines OA, OB, and OC have slope μ, so the viscosity is sdecreasing with increasing *dV/dy*.

Using this definition, we can observe that there are three common types of non-newtonian fluid:

- *Bingham fluids*: Sometimes called *bingham plastics*, resist a small shear stress indefinitely but flow easily under larger shear stresses. One may say that at low stresses the viscosity is infinite and that at higher stresses the viscosity decreases with increasing velocity gradient. Examples are bread dough, tooth paste, jellies, and some slurries.
- *Pseudoplastic fluids:* Show a viscosity that decreases with increasing velocity gradient. Examples are most slurries, muds, polymer solutions, solutions of natural gums, and blood.
- *Dilatant fluids:* Show a viscosity that increases with increasing velocity gradient. They are uncommon, but starch suspensions behave in this way.

The basic unit of viscosity is the poise, where 1P = 1g/ (cm.s) = 0.1 Pa.s = 6.72×10^{-2} lbm/ (ft.s). It is widely used for materials such as high-polymer solutions and molten polymers.

However, it is too large a unit for most common fluids. By sheer coincidence the viscosity of pure water at about 68°F is 0.01 P; for that reason the common unit of viscosity in the United States is the centipoises, where

$$1 \text{ cP} = 0.01 \text{ P} = 0.01 \text{ g/(cm.s)} = 6.72 \times 10^{-4} \text{lb}_m\text{/(ft.s)} = 0.001 \text{Pa.s.}$$

Hence, the viscosity of a fluid, expressed in centipoises, is the same as the ratio of its viscosity to that of water at room temperature.

Kinematic Viscosity

In many engineering problems, viscosity appears only in the relation of viscosity divided by density. Therefore, to save writing, we define,

$$\text{Kinematic Viscosity} = \upsilon = \frac{\mu}{\rho}$$

The most common unit of kinematic viscosity is the centistokes (cSt):

$$1\text{cSt} = \frac{1\text{cP}}{1\text{g/cm}^3} = 1.08 \times 10^{-5}\frac{\text{ft}^2}{\text{s}} = 10^{-6}\frac{\text{m}^2}{\text{s}}$$

(The kinematic viscosity of water at room temperature is 1cSt.) To avoid confusion over which viscosity is being used, some writers refer to the viscosity μ as the absolute viscosity.

FLOW MEASUREMENT

MANOMETER

Pressure (P) is the ratio of the force (F) applied to a surface to the surface area (A).

$$P = F/A$$

A manometer is an device employed to measure pressure. There are a variety of manometer designs. A simple, common design is to seal a length of glass tubing and then bend the glass tube into a U-shape. The glass tube is then filled with a liquid, typically mercury, so that all trapped air is removed from the sealed end of the tube. The glass tube is then positioned with the curved region at the bottom.

After the mercury settles to the bottom of the manometer, a vacuum is produced in the sealed tube. The open tube is connected to the system whose pressure is being measured. In the sealed tube, there is no gas to exert a force on the mercury. In the tube connected to the system, the gas in the system exerts a force on the mercury. The net result is that the column of mercury in the left (sealed) tube is higher than that in the right (unsealed) tube. The difference in the heights of the columns of mercury is a measure of the pressure of gas in the system.

In the example at the left, the top of the left column of mercury corresponds to 875 mm on the scale. The top of the right column of mercury corresponds to 115 mm. The difference in heights is 875 mm - 115 mm = 760. mm, which indicates that the pressure is 760. mm Hg or 760. torr.

This method for measuring pressure led to the use of millimeters of mercury (mm Hg) as a unit of pressure. Today 1 mm Hg is called 1 torr. A pressure of 1 torr or 1 mm Hg is literally the pressure that produces a 1 mm difference in the heights of the two columns of mercury in a manometer. To understand how the height of a column of mercury can be used as a unit of pressure and how the unit of torr is related to the SI unit of pascal (1 Pa = 1 N/ m^2), consider the following mathematical analysis of the behaviour of the manometer. The force exerted by the column of mercury in a tube arises from the gravitational acceleration of the column of mercury. Newton's Second Law provides an expression for this force:

$$F = m\,g$$

In this equation, m is the mass of mercury in the column and $g = 9.80665$ m/sec^2 is the gravitational acceleration. This force is distributed over the cross-sectional area of the column (A). The pressure resulting from the column of mercury is thus,

$$P = \frac{mg}{A}$$

The mass of mercury is given by the product of the density of mercury (d_{Hg}) and the volume of mercury (V). For a cylindrical column of mercury, the volume of mercury is the product of the cross-sectional area and the height of the column (h). These relationships product the following equation.

$$P = \frac{mg}{A} = \frac{d_{Hg}\,V_g}{A} = \frac{d_{Hg}\,AV_g}{A} = d_{Hg}\,Ahg$$

This equation clearly shows that the height of a column of mercury is directly proportional to the pressure exerted by that column of mercury. The difference in heights of the two columns of mercury in a manometer can thus be used to measure the difference in pressures between the two sides of the manometer.

The relation between torr and Pa is also clearly evident. Using d_{Hg} = 13.5951 g cm^{-3}, one finds that 1 torr = 133 Pa or 1 atm = 760 torr = 101 kPa.

Analog Manometers

One of the earliest pressure measuring instruments is still in wide use today because of its inherent accuracy and simplicity of operation. It's the U-tube manometer, which is a U-shaped glass tube partially filled with liquid. This manometer has no moving parts and requires no calibration. Manometry measurements are functions of gravity and the liquid's density, both physical properties that make the U-tube manometer a NIST standard for accuracy.

Digital Manometers

A digital manometer is a microprocessor based instrument that can be stationary or portable. It has communication output capabilities that can be used for process control or transferring measurement data. Due to the network communication capabilities, digital or "smart" manometers are well suited for field measurement and process control.

Pitot tube

A Pitot tube is a pressure measuring instrument used to measure fluid flow velocity, and more specifically, used to determine the airspeed of an aircraft. The Pitot tube was invented by Italian-born French engineer Henri Pitot in the early 1700s, and was modified to its modern form in the mid 1800s by French scientist Henry Darcy.

The basic Pitot tube simply consists of a tube pointing directly into the fluid flow. As this tube contains air, a pressure can be measured as the moving air is brought to rest. This pressure is the stagnation pressure of the air, also known as the total pressure, or sometimes (particularly in aviation circles) the pitot pressure.

The measured stagnation pressure cannot of itself be used to determine the airspeed. However, since Bernoulli's equation states that stagnation pressure = static pressure + dynamic pressure, then the dynamic pressure is simply the difference between the static pressure and the stagnation pressure. The static pressure is generally measured using the static ports on the side of the fuselage.

The dynamic pressure is then determined using a diaphragm inside an enclosed container. If the air on one side of the diaphragm is at the static pressure, and the other at the stagnation pressure, then the deflection of the diaphragm is proportional to the dynamic pressure, which can then be used to determine the indicated airspeed of the aircraft. The diaphragm arrangement is typically contained within the airspeed indicator, which converts the dynamic pressure to an airspeed reading by means of mechanical levers.

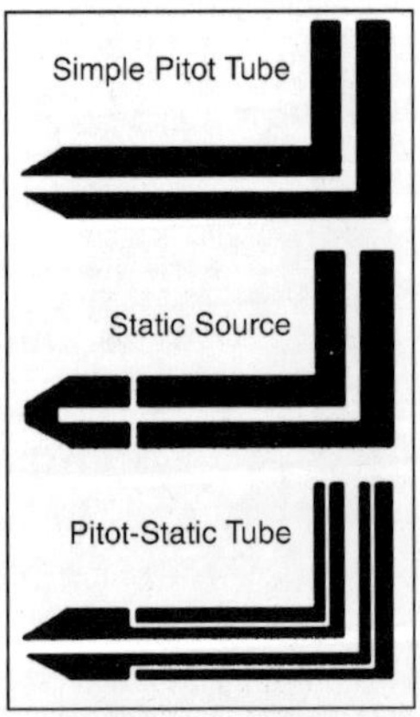

Fig. Types of Pitot Tube

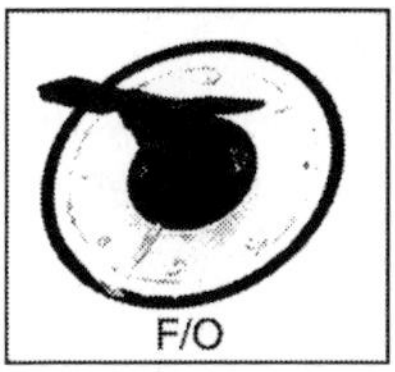

Fig. Airliners have Pitot Probes Protruding from the Cockpit for Measuring Airspeed

Instead of static ports, a Pitot-static tube (also called a Prandtl tube) may be employed, which has a second tube coaxial with the Pitot tube with holes on the sides, outside the direct airflow, to measure the static pressure.

Pitot tubes on aircraft commonly have heating elements to prevent the tube from becoming clogged with ice. The failure of these systems has catastrophic consequences, like in the case of Austral Líneas Aéreas Flight 2553.

Applications in Mining

In mining applications the velocities being measured are those flowing in ducts or tubing where measurements by an anemometer would be difficult to obtain. In these kinds of measurements, the most practical instrument to use is the Pitot tube. The Pitot tube can be inserted through a small hole in the duct with the Pitot connected to a U-tube water gauge or some other differential pressure gauge for determining the velocity inside the duct.

The quantity of air flowing through the duct each minute can then be estimated from:

Volume Flow Rate (cubic feet per minute) = Duct Area (square feet) × Velocity (feet per minute)

Velocity is typically measured in feet per minute in such applications.

Venturi Meter

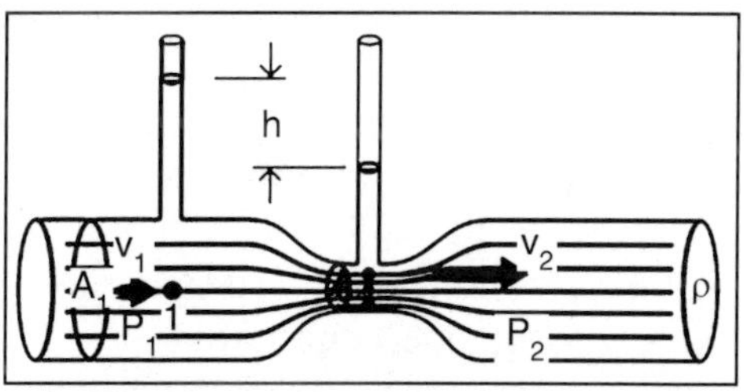

A Venturi meter is shown in a diagram, the pressure in "1" conditions is higher than "2", and the relationship between the fluid speed in "2" and "1" respectively, is the same as for pressure.

The Venturi effect is an example of Bernoulli's principle, in the case of fluid flow through a tube or pipe with a constriction in it. The fluid velocity must increase through the constriction to satisfy the equation of continuity, while its pressure must decrease due to conservation of energy: the gain in kinetic energy is supplied by a drop in pressure or a pressure gradient force. The effect is named after Giovanniee Battista Venturi, (1746–1822), an Italian physicist.

The limiting case of the Venturi effect is choked flow, in which a constriction in a pipe or channel limits the total flow rate through the channel, because the pressure cannot drop below zero in the constriction. Choked flow is used to control the delivery rate of water and other fluids through spigots and other valves.

Referring to the diagram to the right, using Bernoulli's equation in the special case of incompressible fluids (such as the approximation of a water jet), the theoretical pressure drop at the constriction would be given by $(\rho/2)(v_2^2 - v_1^2)$.

Experimental Apparatus

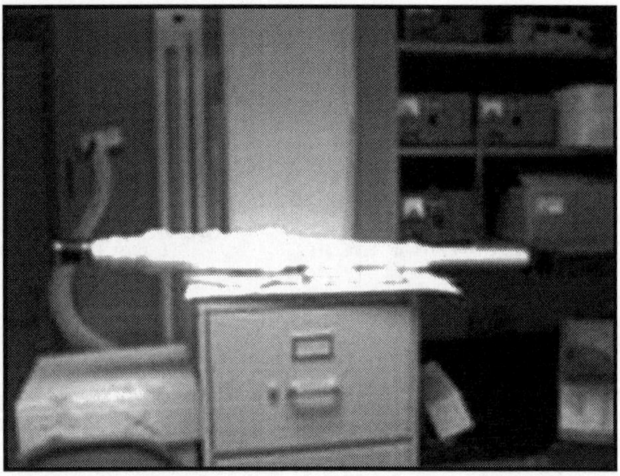

This is a Venturi tube demonstration apparatus built out of PVC pipe and operated with a vacuum pump.

- *Venturi Tubes*: The simplest apparatus, as shown in the photograph and diagram, is a tubular setup known as a Venturi tube or simply a venturi. Fluid flows through a length of pipe of varying diameter. To avoid undue drag, a venturi tube typically has an entry cone of 30 degrees and an exit cone of 5 degrees.

A venturi can also be used to mix a fluid with air. If a pump forces the fluid through a tube connected to a system consisting of a venturi to increase the water speed (the diameter decreases), a short piece of tube with a small hole in it, and last a venturi that decreases speed (so the pipe gets wider again), air will be sucked in through the small hole because of changes in pressure. At the end of the system, a mixture of fluid and air will appear.

- *Orifice plate*: Venturi tubes are more expensive to construct than a simple orifice plate which uses the same principle as a tubular scheme, but the orifice plate causes significantly more permanent energy loss and is less accurate.

In Chronic Aortic Regurgitation, after the initial large stroke volume is released, the Venturi effect draws walls together, transiently obstructing flow causing a Pulsus Bisferiens.

Practical Uses

The Venturi effect is visible in:

- The capillaries of the human circulatory system, where it indicates aortic regurgitation
- Large cities where wind is forced between buildings.
- Inspirators that mix air and flammable gas in barbecues, gas stoves, bunsen burners and airbrushes.
- Water aspirators that produce a partial vacuum using the kinetic energy from the faucet water pressure
- Steam siphon using the kinetic energy from the steam pressure to create a partial vacuum
- Atomizers that disperse perfume or spray paint (*i.e.* From a spray gun)
- Foam firefighting nozzles and extinguishers
- Carburetors that use the effect to suck gasoline into an engine's intake air stream.
- Protein skimmers, a filtration device for saltwater aquaria.
- In automated pool cleaners that use pressure-side water flow to collect sediment and debris.
- The modern day barrel of the clarinet, which uses a reverse taper to speed the air down the tube, enabling better tone, response and intonation.
- Compressed air operated industrial vacuum cleaners
- Venturi scrubbers used to clean flue gas emissions
- Injectors (or sometimes called ejectors) used to add chlorine gas in water treatment chlorination systems

A simple way to demonstrate the Venturi effect is to squeeze and release a flexible hose that is normal shape: the partial vacuum produced in the constriction is sufficient to keep the hose collapsed.

Venturi tubes are also used to measure the speed of a fluid, by measuring pressure changes at different segments of the device. Placing a liquid in a U-shaped tube and connecting the ends of the tubes to both ends of a venturi is all that is needed. When the fluid flows though the venturi the pressure in the two ends of the tube will differ, forcing the liquid to the "low pressure" side.

The amount of that move can be calibrated to the speed of the fluid flow.

Orifice Meters

Good water management requires that flow rates be accurately measured. A simple but economical and accurate device for the measurement of flow rates from pipes discharging to the atmosphere is the orifice meter. The orifice meter is sometimes called a pipe orifice, an end-cap orifice, or more descriptively, a circular orifice weir.

Orifice meters are usually circular orifices placed at the end of a horizontal discharge pipe. Flow rates are calculated from the orifice characteristics and a measure of the pressure behind the orifice. The head (or pressure) on the orifice is measured with a water manometer to obtain a high degree of accuracy (Typically within 2 per cent of true values).

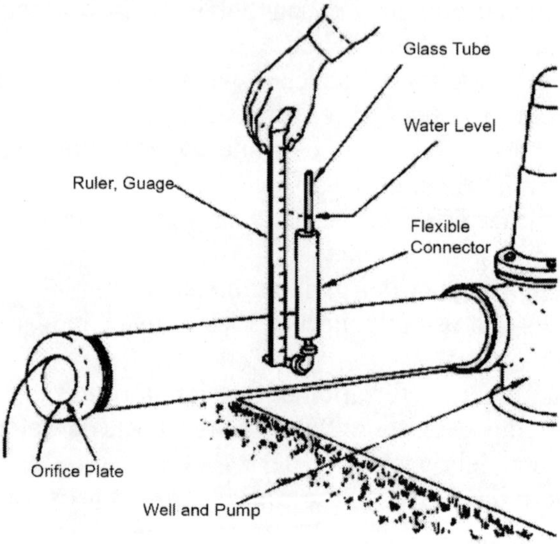

Because of the simplicity of design and the few components of the orifice meter, it can readily be constructed by individuals with only average mechanical skills. An orifice meter can be constructed from a pipe end-cap which has been properly drilled, if the discharge pipe is threaded to allow the end cap to be added for flow measurements or removed after its use.

The cost of the orifice meter is minimal when self-constructed. The orifice meter is very accurate because of the sensitivity of the water manometer, and, because of the absence of moving parts or the need for calibration, it is often used as a means of calibrating other flow-measuring devices.

Details of Construction

The orifice consists of a perfectly round hole in the center of a circular steel plate. The orifice must be cut with clean, square edges. The plate must be 1/16-inch thick around the circumference of the hole. The plate is fastened against the outer end of a level discharge pipe so that the orifice is centered on the pipe. The end of the pipe must be cut square so that the plate will be vertical. A properly machined pipe end-cap may be used as the orifice plate.

The bore of the pipe should be smooth and free of any obstruction that might cause excessive turbulence.

This includes ensuring that there are no elbows, valves or other fittings closer than 6 feet upstream from the orifice. This approach must also be straight and level.

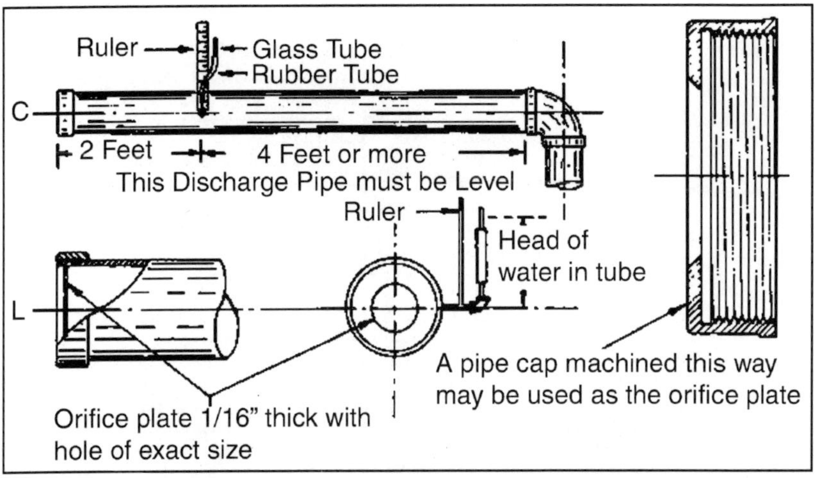

To measure pressure head on the orifice with sufficient accuracy, a glass tube manometer should be used. Exactly 24 inches from the orifice plate, the pipe wall is tapped mid-way between the top and bottom with a 1/8-1/4 inch diameter hole. Burrs inside the pipe caused by the drilling or tapping should be filed to ensure a smooth surface. The glass tube is connected with a short piece of flexible tubing to a pipe nipple threaded into the tapped hole. The glass tube is mounted vertically with a scale set to measure the height of water in the tube above the centertine of the pipe. When water is pumped through the orifice so that it is flowing full, the height of the water in the tube is the pressure head on the orifice.

Orifice Equations

For any given size of orifice discharge pipe, the rate of flow varies in a known manner with the pressure head as measured with the glass tube manometer. Standard tables, such as Table , give the flow in gallons per minute (gpm) for various combinations of orifice diameter and pipe diameter

Discharge through the orifice is computed from equation.

$$Q = 8.02c/\sqrt{h}$$

equation.

where,

Q = Orifice discharge (gpm),

C = Coefficient which varies with the ratio of the orifice Diameter to the pipe diameter, as well as with all of the other factors affecting flows in orifices of this construction,

A = cross sectional area of the orifice (in ^{2}), and

h = head on the orifice measured above its centerline (in).

The value of its coefficient, C, varies with the ratio of the orifice diameter to pipe diameter because of changes in streamlines is flow converges through orifices of various sizes. The value of C is also a function of friction losses through the orifice. Therefore, the values of C given in Figure are only applicable when the orifice meter is constructed as previously described.

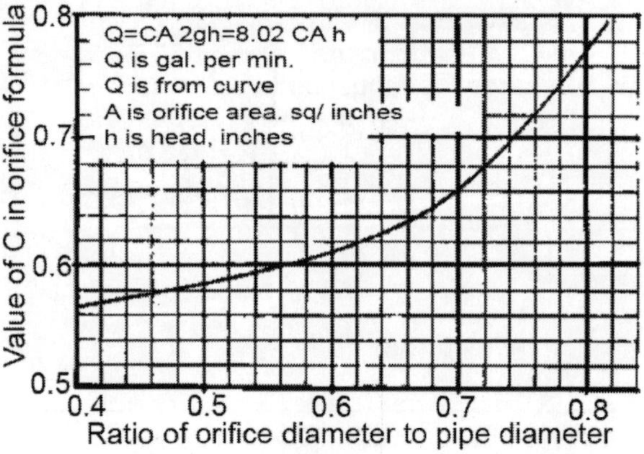

Example Calculations

The use of Equation 1 is illustrated by an example.

The following data apply:

- Orifice constructed as previously described
- Orifice diameter = 4 in
- Discharge pipe inside diameter = 8 in
- Pressure head, h = 30 in

Calculations are performed in 4 steps:

1. Ratio of orifice diameter to pipe diameter = 4 in / 8 in = 0.5
2. From Figure 3, C = 0. 587
3. Cross sectional area of orifice = pi d ^{2}/4 = (3.14)(4 in) ^{2}/4 = 12.56 in ^{2}
4. Q = 8.02 CA h

   Q = 8.02 (0.587) (12.56 in2) 30in

   Q = 324 gpm

Comparing this result with that shown in Table (325 gpm) shows the utility of Equation . The discrepancies between values in Table and those calculated result from inaccuracies in interpolating C values from Figure.

Ensuring Accuracies

The orifice meter provides the capability of measuring flow rates with accuracies within 2 per cent of the actual rates if it is properly constructed and operated. Besides constructing the parts accurately and setting up the device correctly in the field, additional precautions must be taken to obtain accurate results. First, the diameter of the orifice should be less than 0.8 and greater than 0.4 of the inside diameter of the discharge pipe. The value of C changes rapidly as the ratio of diameters increases. Therefore, to ensure greatest accuracy, the ratio should be less than 0.7.

Second, the manometer tube must be completely free of obstruction and air bubbles when measurements of pressure head are made. Air bubbles can be eliminated by lowering the manometer tube to allow water to flow through it.

FLOW MEASUREMENT AND FLUID VISCOSITY

Viscosity is the internal resistance of a fluid to flow. Water has low viscosity, while syrup has high viscosity. When syrup is cold it is very much more viscousthan when it is hot. Viscosity is highly temperature dependent.

ABSOLUTE AND KINEMATIC VISCOSITY

There are two ways to measure the viscosity of a fluid. One is to rotate a cup ordisk in the fluid and measure the rotational speed along with the torque required tokeep it rotating.

The density of the fluid doesn't play any part in the measurement.Viscosity is the torque divided by the speed and divided bythe contacting surfacearea. When measured this way, the viscosity is called "absolute viscosity." Itsunits are centipoise, pascal-seconds, or lb/ft-sec.

Another way to measure viscosity is to allow a certain volume of fluid to drain by gravity out of a container through a capillary tube or some other restriction.The time to drain is directly related to viscosity and is often reported in "seconds."

CONTINUITY EQUATION

Even though a detailed analysis of fluid flow can be extremely difficult, the basic concepts involved in fluid flow problems are fairly straightforward. These basic concepts can be applied in solving fluid flow problems through the use of simplifying assumptions and average values, where appropriate.

Even though this type of analysis would not be sufficient in the engineering design of systems, it is very useful in understanding the operation of systems and predicting the approximate response of fluid systems to changes in operating parameters. The basic principles of fluid flow include three concepts or principles; the first two of which the student has been exposed to in previous manuals. The first is the principle of momentum(leading to equations of fluid forces) which was covered in the manual on Classical Physics.

The second is the conservation of energy (leading to the First Law of Thermodynamics) which was studied in thermodynamics. The third is the conservation of mass (leading to the continuity equation) which will be explained in this module.

PROPERTIES OF FLUIDS

A *fluid* is any substance which flows because its particles are not rigidly attached to one another.

This includes liquids, gases and even some materials which are normally considered solids, such as glass. Essentially, fluids are materials which have no repeating crystalline structure. Several properties of fluids were discussed in the Thermodynamics section of this text.

These included temperature, pressure, mass, specific volume and density. *Temperature* was defined as the relative measure of how hot or cold a material is. It can be used to predict the direction that heat will be transferred. *Pressure* was defined as the force per unit area. Common units for pressure are pounds force per square inch (psi).

Mass was defined as the quantity of matter contained in a body and is to be distinguished from weight, which is measured by the pull of gravity on a body. The *specific volume* of a substance is the volume per unit mass of the substance. Typical units are ft^3/lbm. *Density*, on the other hand, is the mass of a substance per unit volume.

Typical units are lbm/ft^3. Density and specific volume are the inverse of one another. Both density and specific volume are dependant on the temperature and somewhat on the pressure of the fluid. As the temperature of the fluid increases, the density decreases and the specific volume increases.

Since liquids are considered incompressible, an increase in pressure will result in no change in density or specific volume of the liquid. In actuality, liquids can be slightly compressed at high pressures, resulting in a slight increase in density and a slight decrease in specific volume of the liquid.

The properties outlines below are general properties of fluids which are of interest in engineering. The symbol usually used to represent the property is specified together with some typical values in SI units for common fluids. Values under specific conditions (temperature, pressure etc.) can be readily found in many reference books. The dimensions of each unit is also give in the MLT system (see later in the section on dimensional analysis for more details about dimensions.)

Density

The density of a substance is the quantity of matter contained in a unit volume of the substance. It can be expressed in three different ways.

Mass Density

Mass Density, ρ, is defined as the mass of substance per unit volume.
Units: Kilograms per cubic metre, kg/m^3 (or $^{kgm-3}$)
Dimensions: ML^{-3}
Typical values: Water = 1000^{kgm-3}, Mercury = 13546^{kgm-3} Air = 1.23^{kgm-3}, Paraffin Oil = 800^{kgm-3}. (at pressure =1.013 = 1.013×10^{-5} Nm^{-2} and Temperature = 288.15 K.)

Specific Weight

Specific Weight ω, (sometimes, and sometimes known as *specific gravity*) is defined as the weight per unit volume.

or

The force exerted by gravity, g, upon a unit volume of the substance. The Relationship between g and ω can be determined by Newton's 2^{nd}Law, since weight per unit volume = mass per unit volume g,

$$\omega = \rho g$$

Units: Newton's per cubic metre, N/M^3 (or Nm^{-3})
Dimensions: $ML^{-2}T^{-2}$.
Typical values:
Water = 9814 Nm^{-3}, Mercury = 132943 Nm^{-3}, Air = 12.07 Nm^{-3}, Paraffin Oil =7851 Nm^{3}

Relative Density

Relative Density, σ, is defined as the ratio of mass density of a substance to some standard mass density. For solids and liquids this standard mass density is the maximum mass density for water (which occurs at 4°c) at atmospheric pressure.

$$s = \frac{\rho_{sibstamce}}{\rho_{H_2O(\omega 4°c)}}$$

Units: None, since a ratio is a pure number.
Dimensions: 1.
Typical values: Water = 1, Mercury = 13.5, Paraffin Oil =0.8.

Viscosity

Viscosity, m, is the property of a fluid, due to cohesion and interaction between molecules, which offers resistance to sheer deformation. Different fluids deform at different rates under the same shear stress. Fluid with a high viscosity such as syrup, deforms more slowly than fluid with a low viscosity such as water. All fluids are viscous, "Newtonian Fluids" obey the linear relationship given by Newton's law of viscosity which we saw earlier,

$$\tau=\mu\frac{du}{dy},$$

where τ is the shear stress,

Units: Nm^{-2}; kg $m^{-1}s^{-2}$

Dimensions $ML^{-1}T^{-2}$.

$$\frac{du}{dy}$$

is the velocity gradient or rate of shear strain, and has

Units: radianss^{-1}

Dimensions t^{-1}

μ is the "coefficient of dynamic viscosity"-see below.

Coefficient of Dynamic Viscosity

The Coefficient of Dynamic Viscosity, μ, is defined as the shear force, per unit area, (or shear stress τ), required to drag one layer of fluid with unit velocity past another layer a unit distance away.

$$\mu=\tau\frac{du}{dy}=\frac{\text{Force}}{\text{Area}}\frac{\text{Velocity}}{\text{Dis tan ce}}=\frac{\text{Force}\times\text{Time}}{\text{Area}}=\frac{\text{Mass}}{\text{Length}\times\text{Area}}$$

Units: Newton seconds per square metre, Nsm^{-2} or Kilograms per meter per second, $kgm^{-1}\ s^{-1}$. (Although note that μ is often expressed in Poise, P, where 10 P = 1 $kgm^{-1}\ s^{-1}$.)

Typical values: Water $=1.14\times10^{-3}$ Air $=1.78\ \times10^{-5}\ kgm^{-1}s^{-1}$, Mercury $=1.552\ kgm^{-1}\ s^{-1}$,

Paraffin Oil $=1.9\ kgm^{-1}\ s^{-1}$.

Kinematic Viscosity

Kinematic Viscosity, ν, is defined as the ratio of dynamic viscosity to mass density.

$$\nu\frac{\mu}{\rho}$$

Units: square metres per second, m^2s^{-1} (Although note that n is often expressed in Stokes, St, where,

10^4 St = 1 m^2s^{-1}), St = 1 m^2s^{-1})

Dimensions: $L^2\ T^{-1}$.

Typical values:

Water $=1.14\times10^{-6}\ m^2s^{-1}$, Air $=1.46\times10^{-5}\ m^2s^{-1}$, Mercury $=1.145\times10^{-4}\ m^2s^{-1}$,

Paraffin Oil $=2.375\times10^{-3}\ m^2s^{-1}$

BUOYANCY

Buoyancy is defined as the tendency of a body to float or rise when submerged in a fluid. Weall have had numerous opportunities of observing the buoyant effects of a liquid. When we goswimming, our bodies are held up almost entirely by the water. Wood, ice, and cork float onwater. When we lift a rock from a stream bed, it suddenly seems heavier on emerging from thewater. Boats rely on this buoyant force to stay afloat. The amount of this buoyant effect wasfirst computed and stated by the Greek philosopher Archimedes. When a body is placed in afluid, it is buoyed up by a force equal to the weight of the water that it displaces.

2

Fluid Kinematics

GENERAL CONSIDERATIONS

The the most important basic knowledge of mathematics and physics with respect to fluid mechanics. This knowledge is needed to describe fluid flows or derive and construct basic equations of fluid mechanics in order to solve flow problems.

Here it is important to know that fluid mechanics is primarily interested in the velocity field $U_j(x_i, t)$ at initial and boundary conditions, and in the accompanying pressure field $P(x_i, t)$, *i.e.* fluid mechanics tries to describe flow processes in field variables. This representation results in "Eulerian presentation" of fluid flows.

This is best suited for the solution of flow problems and is thus applied in experimental, analytical and numerical fluid mechanics. Thanks to the introduction of field quantities for the thermodynamic properties of a fluid, like *e.g.* the pressure $P(x_i, t)$, the temperature $T(x_i, t)$, the density $\rho(x_i, t)$, the internal energy $e(x_i, t)$ as well as for the molecular transport quantities, as the dynamic vis-cosity $\mu(x_i, t)$, the heat conductivity $\lambda(x_i, t)$ and the diffusion coefficients $D(x_i, t)$ so that a complete presentation of fluid mechanics is possible. With the inclusion of diffusive transport quantities, *i.e.* the molecular heat trans-port $q_i(x_i, t)$, the molecular mass transport $m_i(x_i, t)$ the molecular momen-tum transport $t_{ij}(x_i, t)$, it is possible to formulate the conservation laws for mass, momentum and energy for general application.

The basic equations of fluid mechanics can thus be formulated locally, and hold for all flow problems in the same form. The differences in the solutions result from the different initial and boundary conditions that define the actual flow problems which enter into the solutions by the integration of the locally formulated basic quations.

Experience shows that the derivation of basic equations of fluid mechanics can be achieved in the easiest way if considerations are carried for fluid ele-ments, *i.e.* by employing the "Lagrangian consideration" for the derivation of equations. The "Lagrange considerations" assumes that a fluid can be split up at a fixed time $t = 0$, in "marked elements" with the mass $\delta m_{\Re}$, the pres-sure $P_{\Re}$, the temperature $T_{\Re}$, the density $\rho_{\Re}$, the internal energy $e_{\Re}$ etc. The element with the index $\Re$ possesses also a velocity $(U_j)_{\Re}$, which is defined as Lagrangian velocity and which is always linked to the fluid element marked Basics of Fluid Kinema-tics once with as well as to all other quantities labelled with $\Re$. In fluid mechan-ics, these are also designated as substantial quantities and always employed to derive the basic laws of fluid mechanics in an easily comprehensible way. As the following considerations will show, the basic knowledge of mechanics gained in physics can be transferred in the most simple way into fluid mechanics by deriving the basic equations for fluid elements by introducing the Lagrange considerations.

SUBSTANTIAL DERIVATIVES

While one defines a generally as a substantial quantity, the derivation of fluid mechanics equations often require the total differential da to be employed.

$$d\alpha_{\Re} = \frac{\partial \alpha}{\partial t} dt + \frac{\partial \alpha}{\partial x_1}(dx_1)_{\Re} + \frac{\partial \alpha}{\partial x_2}(dx_2)_{\Re} + \frac{\partial \alpha}{\partial x_2}(dx_3)_{\Re}$$

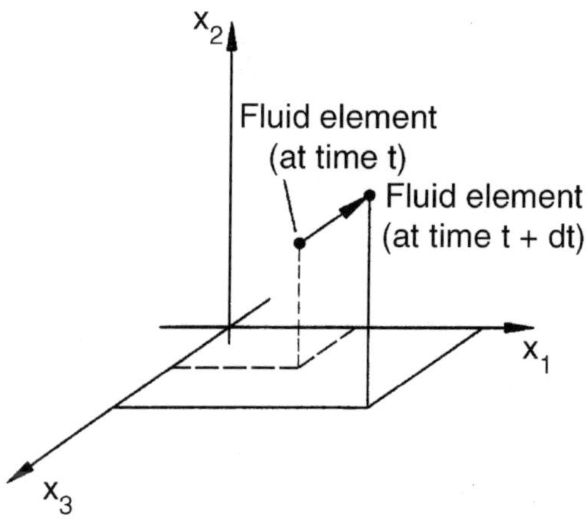

Fig. Motion of a fluid element in space

The fluid element motion in space can be described as follows:

$(dx_1) \quad = (U_1)\, dt = U_1 dt$

$(dx_2) \quad = (U_2)\, dt = U_2 dt$

$(dx_3) \quad = (U_3)\, dt = U_3 dt$

The transition of substational velocities $(U_i)_{\Re}$ to the field quantities U_i in equation is permissible, as at the time $t(x_i) = x_i(t)$ and thus $(U_i)(t) = U_i(x_i, t)$. Therefore generally it holds:

$$d\alpha_{\Re} = \frac{\partial \alpha}{\partial t} dt + \frac{\partial \alpha}{\partial x_1} U_1 dt \frac{\partial \alpha}{\partial x_2} U_2 dt + \frac{\partial \alpha}{\partial x_3} U_3 dt$$

and for the substantial time derivative with $\alpha_{\Re}(t) = \alpha(x_i, t)$ when $x_{\Re} = x_i$ at the time t it holds:

$$\frac{d\alpha_{\Re}}{dt} = \frac{\partial \alpha}{\partial t} + U_i \frac{\partial \alpha}{\partial x_i} =: \frac{D\alpha}{Dt}$$

where (Dα/Dt) is the substantial time derivative of the field quantity $\alpha(x_i, t)$ with respect to time and the operator:

$$\frac{D}{Dt} := \frac{\partial}{\partial t} + U_i \frac{\partial}{\partial x_i}$$

indicates how the substantial derivative of a field variable is to be calculated. The operator D/D*t* may only be applied to field quantities. When one applies D/D*t* to the velocity field $U_j(x_i, t)$, the substantial acceleration results, *i.e.* the local acceleration which a fluid element experiences in a flow field at a point x_i at time t where $U_j(x_i, t)$ exists.

$$\frac{DU_j}{Dt} = \frac{\partial U_j}{\partial t} + U_i \frac{\partial U_j}{\partial x_i}$$

The substantial derivative plays an important role in the derivation of the momentum equation of fluid mechanics in Euler variables. In the acceleration term in Euler variables, four partial derivatives occur per momentum direction $j = 1, 2, 3$, one time derivative and three derivatives with respect to the space coordinates x_1, x_2 and x_3 *i.e.* the spatial derivatives $(\partial U_j / \partial x_i)$ multiplied by U_i occur in the substantial acceleration. These non-linear terms in the resulting momentum equations lead to mathematical complications when flow problems are to be solved.

They prevent the application of the superposition principle of solutions, result in solution bifurcations, *i.e.* in multiple solutions for equal initial and boundary conditions and in coupled velocity fluctuations, *e.g.* in turbulent flows.

The treatment of these non-linear terms is considered in the presentations of this book. It is important that their significance is understood in detail as part of the acceleration term of fluid elements. It is important to realize that not only the temporal changes of the velocity field that lead to accelerations of fluid elements, but also the motion of a fluid element in a non-uniform velocity field.

THE EQUATION OF CONTINUITY

Having introduced our continuum, or hydrodynamic, variables, we now need to construct equations of motion for these variables. We do this by appealing to conservation laws, such as conservation of mass, momentum, and energy.

In this section we will focus on the conservation of mass and derive the *equation of continuity* for the mass density. Consider an arbitrary, fixed volume V, inside the fluid,

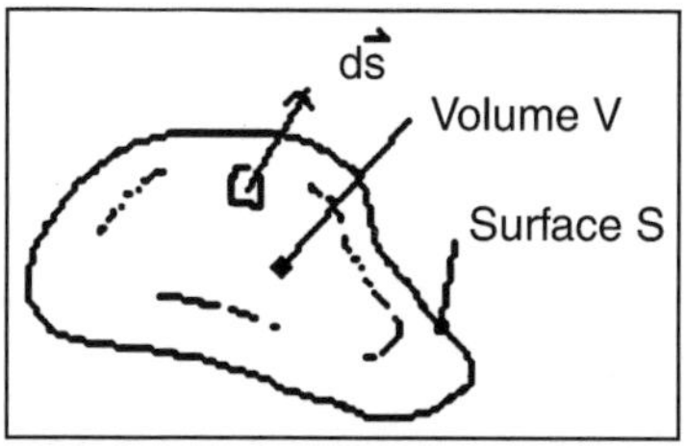

Fig. Fluid Volume Used for the Derivation of the Continuity Equation.

Fluid is entering and leaving V by crossing the surface S. If we let dS be an element of surface area which points outward, normal to the surface, and v be the fluid velocity at the surface, then it is the component of v in the direction of dS which is responsible for transporting mass out of V. Therefore, the mass flux (mass per unit time) transported through the infinitesimal surface area dS . Integrating this over the entire surface S, we have the net rate of change of mass in the volume V is

$$\text{rate of mass change in V} = -\int \rho \mathrm{v} \cdot \mathrm{dS} = -\int_{V} \nabla \cdot (\rho \mathrm{v}) \mathrm{dV},$$

where the minus sign has been added since the mass is decreasing, and the last equality follows from the divergence theorem. Now the total mass in the volume V is,

$$\text{Total Mass in V} = \int_{V} \rho \mathrm{dV},$$

and the rate at which this mass is changing with time is,

$$\text{rate of mass change in V} = \frac{\mathrm{d}}{\mathrm{dt}} \int_{V} \rho \mathrm{dV} = \int_{V} \frac{\partial \rho}{\partial \mathrm{t}} \mathrm{dV}.$$

Since mass is conserved in an arbitrarily small volume V, the integrands of Eqs. and we have,

$$\frac{\partial \rho}{\partial \mathrm{t}} + \nabla \cdot (\rho \mathrm{V}) = 0.$$

This is the equation of continuity, which is a mathematical statement of conservation of mass.

INCOMPRESSIBLE FLOW

Many liquids are hard to compress (*i.e.*, their volumes or densities don't change much when you squeeze them), so that the density ρ is essentially a constant. For such an incompressible fluid the equation of continuity simplifies to

$$\nabla \cdot v = 0 \text{ (incompressible)},$$

so that the velocity field v is a solenoidal vector. This condition is analogous to the condition $\nabla \cdot B =$ which we encounter in electromagnetism—the magnetic field has zero divergence.

Incompressible flows are much easier to treat theoretically than compressible flows, so this is a very useful simplifying assumption. You might worry about applying it to gases, but it turns out that even there it often works well. In fact, if U is some characteristic flow velocity, and c is the speed of sound in the fluid, then the fluid can be treated as incompressible whenever $U << c$. This is often expressed in terms of the Mach number Ma, defined as

$$Ma = U / c;$$

then if $Ma << 1$ the fluid can be treated as incompressible. For instance, the speed of sound in air is 330 m/s (740 mi/h); therefore, fluid flow around people, cars, trains, as well as fluid flow in hurricanes and tornadoes, all satisfy the incompressibility condition. Air flow around commercial airliners ($Ma = 0.7$) violates this condition, as does supersonic flight.

STREAMLINES AND STREAMTUBES

In problems involving electric or magnetic fields we visualize the fields by introducing field lines (often called lines of force), which have the properties that they are everywhere tangent to the field vectors, and the density of the field lines is proportional to the magnitude (strength) of the field. We can use the same method to visualize the velocity field v by introducing streamlines, which have the following properties:

- A tangent to a streamline at a point is in the direction of the fluid velocity at that point;
- The density of streamlines in the vicinity of a point is proportional to the magnitude of the velocity at that point;
- The streamlines cannot intersect except at a point of zero velocity, otherwise the velocity would not be uniquely determined at that point.

A *streamtube* is a tubular region of fluid surrounded by streamlines. Since streamlines don't intersect, the same streamlines pass through a streamtube at all points along its length. Let's take two cross-sections of a streamtube, with cross-sectional areas A_1 and A_2

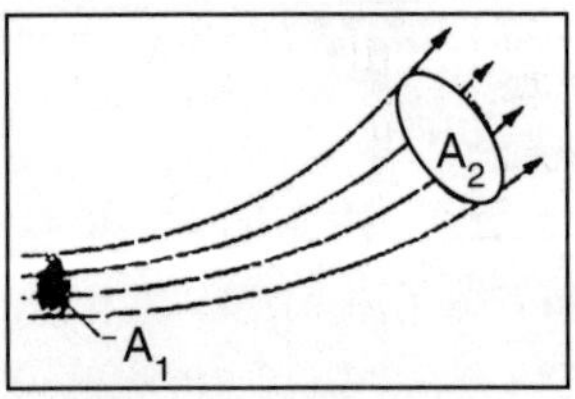

Fig. A Streamtube in a Fluid.

If these areas are made small enough then the fluid velocities across the cross-sections will be constant; let's call the speeds v_1 and v_2 (the directions of the velocities being perpendicular to the cross-sections). The rate at which mass is entering the streamtube is $\rho_1 v_1 A_1$; the rate at which it is leaving is $\rho_2 v_2 A_2$. If the mass inside the streamtube is not changing with time, so that the fluid is either incompressible or in steady-state, we have,

$$\rho_1 v_1 A_1 = \rho_2 v_2 A_2.$$

If, in addition, the fluid is incompressible, then $\rho_1 = \rho_2$, and we have

$$v_1 A_1 = v_2 A_2 A,$$

so that the fluid speed increases when the cross-sectional area of the streamtube decreases. If this seems counter-intuitive, the following analogy might help. You drive down a four lane highway, and due to construction one lane on your side has been closed. The traffic crawls towards the point where the lane has been closed, with cars haphazardly merging into your lane. You finally hit the single lane, and whiz through the construction area. The cross-sectional area has decreased, and your speed has increased.

CIRCULATION, VORTICITY, AND VORTICES

Fluids will invariably swirl around, and it is often useful to characterize the amount of swirling. This is done by introducing the *circulation* Γ, defined as,

$$\Gamma = \oint_C \mathrm{v} \cdot d\mathrm{l},$$

with the line integral taken is around a closed circuit *C* in a counter-clockwise sense. By using Stokes' theorem we may write this as a surface integral,

$$\oint_C \mathrm{v} \cdot d\mathrm{l} = \int_S (\nabla \times \mathrm{v}) \cdot d\mathrm{S}$$
$$= \int_S \mathrm{w} \cdot d\mathrm{S},$$

where the *vorticity* *w* of the fluid is defined as,

$$\mathrm{w} = \nabla \times \mathrm{v}.$$

If the vorticity of the fluid is zero, we say that the fluid is irrotational :

$$\nabla \times \mathrm{v} = 0 \qquad (\text{irrotational}).$$

Already mentioned that there are some useful analogies between magnetostatics and incompressible fluid flow. To transform the magnetostatics expressions into those for fluid flow, we make the substitutions $B/\mu_0 \rightarrow v$, $I \rightarrow \Gamma$, , and $\mu_0 \rightarrow \rho$, (where here ρ is the mass density of the fluid). If a fluid is both incompressible *and* irrotational, then $\nabla \times v = 0$ and $\nabla \cdot v = 0$, which are the same as the equations of electrostatics and magnetostatics in free space (*i.e.*, in the absence of charges and currents).

Magnetostatics

Incompressible fluid flow,

$$\nabla \cdot B = 0$$

$$\nabla \cdot v = 0$$

$$\int B \cdot dl = \mu_0 I \text{ (Ampére's Law)}$$

$$\int v \cdot dl = \Gamma \text{ (circulation)}$$

$$E_B = B^2 / 2_{\mu_0} \text{ (energy density of the magnetic field)}$$

$$E = \frac{1}{2}\rho v^2 \text{ (kinetic energy density of a fluid)}$$

Analogies between incompressible fluid flow and magnetostatics.

As an example of fluid flow with vorticity, we can consider a single *vortex* , which would be produced when draining a bathtub, and which is illustrated in Fig

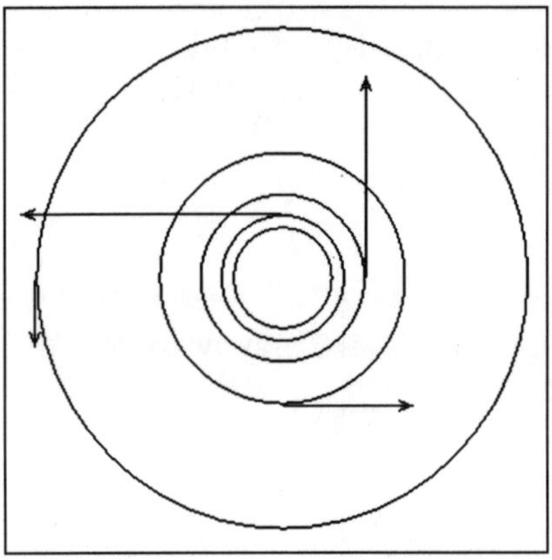

Fig. Velocity Field of a Vortex. Flow Lines are Drawn for Equal Increments of the Speed, as Shown by Arrows.

In cylindrical coordinates (r, θ, z) there is only one component of the velocity field, v_θ. In calculating the circulation, the line element $dl = rd\theta\hat{\theta}$, so that $v \cdot dl = v_\theta rd\theta$. If the circulation is independent of the integration path, then we must have $v_\theta = C/r$, with *C* a constant. The circulation is then,

$$\Gamma = \int_0^{2\pi} \left(\frac{C}{r}\right) rd\theta = 2\pi C,$$

so that $C = \Gamma / 2\pi$. Therefore, the velocity field of a vortex is,

$$v(r) = \frac{\Gamma}{2\pi r}\hat{\theta} \quad (\text{vortex}).$$

By making the substitutions suggested above, we can use this result to find the magnetic field produced by a wire carrying a current *I*:

$$B(r) = \frac{\mu_0 I}{2\pi r}\hat{\theta} \quad (\text{current} - \text{carrying wire}).$$

CAPILLARITY

WHAT IS CAPILLARY ACTION?

Capillary action is a physical effect caused by the interactions of a liquid with the walls of a thin tube. The capillary effect is a function of the ability of the liquid to wet a particular material.

The liquid for which this effect is most commonly seen is water, because water is capable of strong surface interactions and because water is ubiquitous.

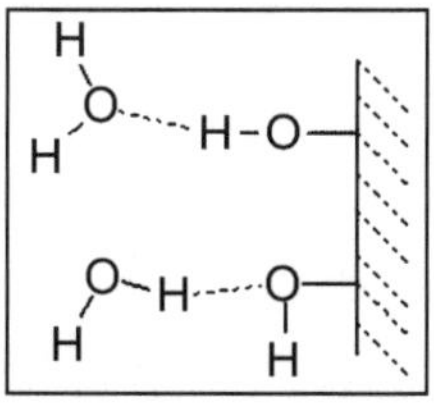

Water climbs up a thin glass tube because of the strong hydrogen-bonding interactions between the water and the oxygens (and terminal hydrogens) at the surface of the glass (SiO_2; surface oxygens are typically bonded to hydrogen). The energetic gain from the new intermolecular interactions must be balanced against gravity, which attempts to pull the liquid back down. Therefore, the narrower the tube, the higher the liquid will climb, because a narrow column of liquid weighs less than a thick one.

One gets the same effect for pores and narrow spaces in any hydrophilic object. While I have not been able to find any information on the Web, it is my suspicion that a narrow tube coated with a hydrophobic substance (such as Teflon) should not show a capillary effect.

Capillary Action

Capillary action is the result of adhesion and surface tension. Adhesion of water to the walls of a vessel will cause an upward force on the liquid at the edges and result in a meniscus which turns upward. The surface tension acts to hold the surface intact, so instead of just the edges moving upward, the whole liquid surface is dragged upward.

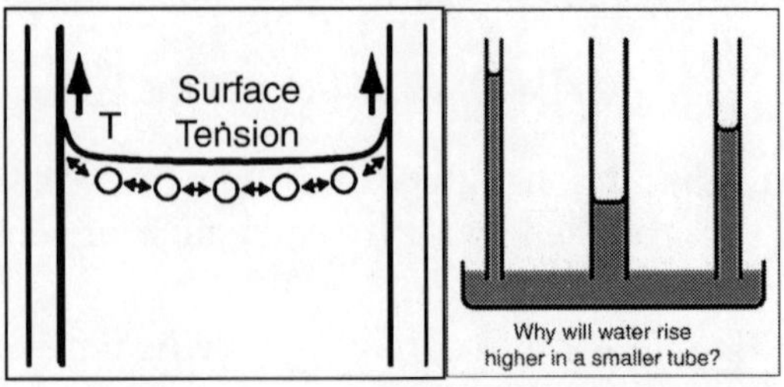

Capillary action, capillarity, capillary motion, or wicking is the ability of a substance to draw another substance into it. The standard reference is to a tube in plants but can be seen readily with porous paper. It occurs when the adhesive intermolecular forces between the liquid and a substance are stronger than the cohesive intermolecular forces inside the liquid. The effect causes a concave meniscus to form where the substance is touching a vertical surface. The same effect is what causes porous materials such as sponges to soak up liquids.

A common apparatus used to demonstrate capillary action is the *capillary tube*. When the lower end of a vertical glass tube is placed in a liquid such as water, a concave meniscus forms. Surface tension pulls the liquid column up until there is a sufficient weight of liquid for gravitational forces to overcome the intermolecular forces. The weight of the liquid column is proportional to the square of the tube's diameter, but the contact length (around the edge) between the liquid and the tube is proportional only to the diameter of the tube, so a narrow tube will draw a liquid column higher than a wide tube. For example, a glass capillary tube 0.5 mm in diameter will lift a theoretical 2.8 cm column of water. Actual observations show shorter total distances.

With some pairs of materials, such as mercury and glass, the interatomic forces within the liquid exceed those between the solid and the liquid, so a convex meniscus forms and capillary action works in reverse.

Examples

In hydrology, capillary action describes the attraction of water molecules to soil particles. Capillary action is responsible for moving groundwater from wet areas of the soil to dry areas. Differences in soil matric potential (ψ_m) drive capillary action in soil.

Capillary action is also essential for the drainage of constantly produced tear fluid from the eye, two canalicula of tiny diameter are present in the inner corner of the eyelid, also called the lacrimal ducts; their openings can be seen with the naked eye within the lacrimal sacs when the eyelids are everted.

Paper towels adsorb liquid through capillary action, allowing a fluid to be transferred from a surface to the towel. The small pores of a sponge act as small capillaries, causing it to adsorb a comparatively large amount of fluid.

Some modern sport and exercise fabrics, such as Coolmax, use capillary action to "wick" sweat away from the skin. These are often referred to as wicking fabrics, presumably after the capillary properties of a candle wick.

Chemists utilize capillary action in thin layer chromatography, in which a solvent moves vertically up a plate via capillary action. Dissolved solutes travel with the solvent at various speeds depending on their polarity.

Capillary Action Calculation

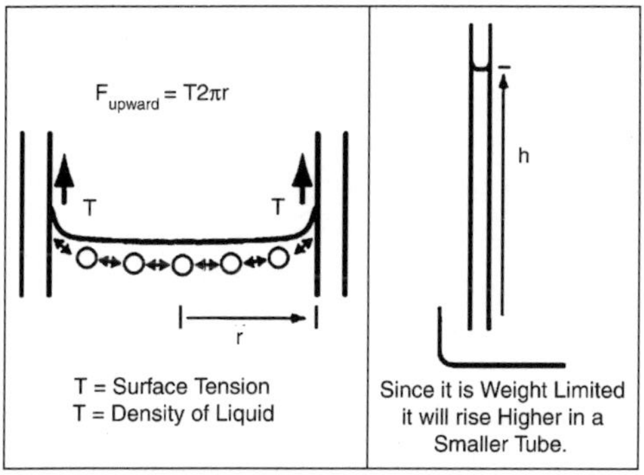

The height h to which Capillary action will lift water depends upon the weight of water which the surface tension will lift:

$$T2\pi r = \rho g\left(h\pi r^2\right)$$

The height to which the liquid can be lifted is given by,

$$h = \frac{2T}{\rho r g}$$

Discussion: With notes on the dimension in SI units, the height *h* of a liquid column (m) is given by,

$$h = \frac{2\gamma \cos\theta}{\rho r g}$$

...where:

γ = The liquid-air surface tension (J/m^2 or N/m)
θ = Contact angle
ρ = Density of liquid (kg/m^3)
g = Acceleration due to gravity (m/s^2)
r = Radius of tube (m)

For a water-filled glass tube in air at sea level,

γ = 0.0728 J/m^2 at 20 °C
θ = 20° (0.35 rad)
∇ = 1000 kg/m^3
g = 9.8 m/s^2

...and so the height of the water column is given by:

$$h \approx \frac{1.4 \times 10^{-5}}{R}.$$

Thus for a 2 m wide (1 m radius) tube, the water would rise an unnoticeable 0.014 mm. However, for a 2 cm wide tube, the water would rise 1.4 mm and for tube with radius 0.1 mm, the water would rise 14 cm (about 6 inches).

Capillary Wave

A capillary wave is a wave travelling along the interface between two fluids, whose dynamics are dominated by the effects of surface tension. Capillary waves are common in nature and the home and are often referred to as ripples. The wavelength of capillary waves is typically less than about a centimeter.

The dispersion relation for capillary waves is,

$$w^2 = \frac{\sigma}{\rho + \rho'} k^3$$

where ω is the frequency, σ the surface tension, ρ the density of the heavier fluid, ρ' the density of the lighter fluid and k the wavenumber. The wavelength is,

$$\lambda = \frac{2\pi}{k}$$

The waves with large wavelengths are generally also affected by gravity and are then called gravity-capillary waves. Their dispersion relation reads, for infinite depth of the two fluids,

$$w^2 = \frac{g(\rho - \rho')}{\rho + \rho'} k + \frac{\sigma}{\rho - \rho'} k^3$$

where ω is the frequency, g the acceleration due to gravity, σ the surface tension, ρ the density and k the wavenumber. This class of waves involves ocean surface waves.

In water on earth this is observed: Gravity waves have a group velocity half the phase velocity.

Following a single wave in a group one can see the wave appearing at the back of the group, growing and finally disappearing at the front of the group.

Therefore an interesting and common situation occurs when the dispersion caused by gravity cancels out the dispersion due to the capillary effect. At wavelength around 2 cm the capillary effect causes group velocity to equal phase velocity. The dispersion is zero, and a wave ridge can travel for long distances.

Shorter (*i.e.* 2 mm) ripples do the opposite: the wave appears at the front of the group, growing and finally disappearing at the back of the group.

Capillary Pressure

In fluid statics, capillary pressure is the difference in pressure across the interface between two immiscible fluids. The pressure difference is proportional to the surface tension, ã, and inversely proportional to the effective radius, *r*, of the interface, it also depends on the wetting angle, θ, of the liquid on the surface of the capillary.

$$Pc = \frac{2\gamma \cos\theta}{r}$$

The capillary pressure p_c is defined as,

$$p_c = p_g - p_w$$

for the case of one phase being gas with pressure p_g and the other phase being water with pressure p_w.

The equation for capillary pressure is only valid under capillary equilibrium, which means that there can not be any flowing phases.

Capillary Number

In fluid dynamics, the capillary number represents the relative effect of viscous forces versus surface tension acting across an interface between a liquid and a gas, or between two immiscible liquids. It is defined as

$$Ca \stackrel{\text{def}}{=} \frac{\mu V}{\sigma}$$

where μ is the viscosity of the liquid, *v* is a characteristic velocity and σ is the surface or interfacial tension between the two fluid phases.

For low capillary numbers (a rule of thumb says less than 10^{-5}), flow in porous media is dominated by capillary forces.

VAPOUR PRESSURE

Vapour pressure is the pressure of a vapor in equilibrium with its non-vapor phases.

All solids and liquids have a tendency to evaporate to a gaseous form, and all gases have a tendency to condense back. At any given temperature, for a particular substance, there is a partial pressure at which the gas of that substance is in dynamic equilibrium with its liquid or solid forms. This is the vapor pressure of that substance at that temperature.

In meteorology, the term *vapor pressure* is used to mean the partial pressure of water vapor in the atmosphere, even if it is not equilibrium, and the equilibrium vapor pressure is specified as such. Meteorologists also use the term saturation vapor pressure to refer to the equilibrium vapor pressure of water or brine above a flat surface, to distinguish it from equilibrium vapor pressure which takes into account the shape and size of water droplets and particulates in the atmosphere.

This article is about our first definition of vapor pressure, or what meteorologists would call equilibrium vapor pressure.

Equilibrium vapor pressure is an indication of a liquid's evaporation rate. It relates to the tendency of molecules and atoms to escape from a liquid or a solid. A substance with a high vapor pressure at normal temperatures is often referred to as *volatile*. The higher the vapor pressure of a material at a given temperature, the lower the boiling point.

The vapor pressure of any substance increases non-linearly with temperature according to the Clausius-Clapeyron relation. The boiling point of a liquid is the temperature where the vapor pressure equals the ambient atmospheric pressure. At the boiling temperature, the vapor pressure becomes sufficient to overcome atmospheric pressure and lift the liquid to form bubbles inside the bulk of the substance.

Units of Vapor Pressure

The most common unit for vapor pressure is the torr. 1 torr = 1 mm Hg (one millimeter of mercury). The international unit for pressure is: 1 pascal = a force of 1 newton per square meter = 10 dyn/cm^2 = 0.01 mbar= 0.0075 mmHg = 0.00000969 atm= 0.0003545414 psi .

Equilibrium Vapor Pressure of Solids

Equilibrium vapor pressure can be defined as the pressure reached when a condensed phase is in equilibrium with its own vapor. In the case of an equilibrium solid, such as a crystal, this can be defined as the pressure when the rate of sublimation of a solid matches the rate of deposition of its vapor phase. For most solids this pressure is very low, but some notable exceptions are naphthalene, dry ice (the vapor pressure of dry ice is 5.73 MPa (831 psi, 56.5 atm) at 20 degrees Celsius, meaning it will cause most non-ventilated containers to explode if sealed inside), and ice. Ice will still continue to disappear even though the ambient temperature is below the freezing point of water. All solid materials have a vapor pressure. However, due to their often extremely low values, measurement can be rather difficult. Typical techniques include the use of thermogravimetry and gas transpiration.

Relation between Solid and Liquid Vapor Pressures

It may be noted that the vapor pressure of a substance in liquid form is usually different from the vapor pressure of the same substance in solid form. If the temperature is such that the vapor pressure of the liquid is higher than that of the solid, liquid will vaporize but vapor will condense to a solid, *i.e.* the liquid is freezing. If the temperature is such that the vapor pressure of the liquid is lower than that of the solid, solid will vaporize but vapor will condense to a liquid, *i.e.* the solid is melting. At the temperature that equalizes the two vapor pressures, an equilibrium exists between solid and liquid phases. This temperature is referred to as the melting point.

Water Vapor Pressure

Graph of water vapor pressure versus temperature. Note that at the boiling point of 100°C, the vapor pressure equals the standard atmospheric pressure of 760 Torr.

Water, like all liquids, starts to boil when its vapor pressure reaches its surrounding pressure. At higher elevations the atmospheric pressure is lower and water will boil at a lower temperature. The boiling temperature of water for pressures around 100 kPa can be approximated by,

$$T_v = 100 + 0.0002772 \cdot (p - 101000) - 1.24 \cdot 10^{-9} \cdot (p - 101000)^2$$

where the temperature T_v is in degrees Celsius and the pressure p is in pascals. One gets the vapor pressure by solving this equation for p.

In meteorology, the international standard for the vapour pressure of water over a flat surface is given by the Goff-Gratch equation.

Partial Pressures

Raoult's law gives an approximation to the vapor pressure of mixtures of liquids. It states that the activity (pressure or fugacity) of a single-phase mixture is equal to the mole-fraction-weighted sum of the components' vapor pressures:

$$P_{tot} = \sum_i P_i X_i$$

where p is vapor pressure, i is a component index, and X is a mole fraction. The term $p_i X_i$ is the vapor pressure of component i in the mixture. Raoult's Law is applicable only to non-electrolytes (uncharged species); it is most appropriate for non-polar molecules with only weak intermolecular attractions (such as London forces).

Systems that have vapor pressures higher than indicated by the above formula are said to have positive deviations. Such a deviation suggests weaker intermolecular attraction than in the pure components, so that the molecules can be thought of as being "held in" the liquid phase less strongly than in the pure liquid. An example is the azeotrope of approximately 95 per cent ethanol and water. Because the azeotrope's vapor pressure is higher than predicted by Raoult's law, it boils at a temperature below that of either pure component.

There are also systems with negative deviations that have vapor pressures that are lower than expected. Such a deviation is evidence for stronger intermolecular attraction between the constituents of the mixture than exists in the pure components.

Thus, the molecules are "held in" the liquid more strongly when a second molecule is present. An example is a mixture of trichloromethane (chloroform) and 2-propanone (acetone), which boils above the boiling point of either pure component.

EXAMPLES OF VAPOR PRESSURES

Table. Material Properties

Gas	Vapor pressure (bar)	Vapor pressure (mmhg)	Temperature
Helium	1	750	@ -269.15 °C
Propane	22	16500	@ 55 °C
Butane	2.2	1650	@ 20 °C
Carbonyl sulfide	12.55	9412	@ 25 °C
Acetaldehyde	0.987	740	@ 20 °C
Freon 113	0.379	284	@ 20 °C
Methyl isobutyl ketone	0.02648	19.86	@ 25 °C
Tungsten	0.001	0.75	@ 3203 °C

COMPRESSIBILITY

This article is about thermodynamics and fluid mechanics. For other uses, see Compression.

- *Specific heat*:

$$c = \frac{T}{N}\left(\frac{\partial S}{\partial T}\right)$$

- *Compressibility*:

$$\beta = -\frac{1}{V}\left(\frac{\partial V}{\partial p}\right)$$

- *Thermal expansion*:

$$\alpha = \frac{1}{V}\left(\frac{\partial V}{\partial T}\right)$$

In thermodynamics and fluid mechanics, compressibility is a measure of the relative volume change of a fluid or solid as a response to a pressure (or mean stress) change.

$$\beta = -\frac{1}{V}\frac{\partial V}{\partial p}$$

where V is volume and p is pressure. The above statement is incomplete, because for any object or system the magnitude of the compressibility depends strongly on whether the process is adiabatic or isothermal. Accordingly we define the isothermal compressibility as:

$$\beta_T = -\frac{1}{V}\left(\frac{\partial V}{\partial p}\right)_T$$

where the subscript T indicates that the partial differential is to be taken at constant temperature.

The adiabatic compressibility as:

$$\beta_s = -\frac{1}{V}\left(\frac{\partial V}{\partial p}\right)_S$$

where S is entropy. For a solid, the distinction between the two is usually negligible.

The inverse of the compressibility is called the bulk modulus, often denoted K (sometimes B). That page also contains some examples for different materials.

Importance of Compressibility in Fluid Dynamics

Compressibility is an important factor in aerodynamics. At low speeds, the compressibility of air is not significant in relation to aircraft design, but as the airflow nears and exceeds the speed of sound, a host of new aerodynamic effects become important in the design of aircraft. These effects, often several of them at a time, made it very difficult for World War II era aircraft to reach speeds much beyond 800 km/h (500mph).

Some of the minor effects include changes to the airflow that lead to problems in control. For instance, the P-38 Lightning with its thick high-lift wing had a particular problem in high-speed dives that led to a nose-down condition. Pilots would enter dives, and then find that they could no longer control the plane, which continued to nose over until it crashed. Adding a "dive flap" beneath the wing altered the center of pressure distribution so that the wing would not lose its lift. This fixed the problem.

A similar problem affected some models of the Supermarine Spitfire. At high speeds the ailerons could apply more torque than the Spitfire's thin wings could handle, and the entire wing would twist in the opposite direction. This meant that the plane would roll in the direction opposite to that which the pilot intended, and led to a number of accidents. Earlier models weren't fast enough for this to be a problem, and so it wasn't noticed until later model Spitfires like the Mk.IX started to appear. This was mitigated by adding considerable torsional rigidity to the wings, and was wholly cured when the Mk.XIV was introduced.

The Messerschmitt Bf 109 and Mitsubishi Zero had the exact opposite problem in which the controls became ineffective. At higher speeds the pilot simply couldn't move the controls because there was too much airflow over the control surfaces. The planes would become difficult to maneuver, and at high enough speeds less maneuverable aircraft could out-turn them.

Finally, another common problem that fits into this category is flutter. At some speeds the airflow over the control surfaces will become turbulent, and the controls will start to flutter. If the speed of the fluttering is close to a harmonic of the control's movement, the resonance could break the control off completely. This was a serious problem on the Zero. When problems with poor control at high speed were first encountered, they were addressed by designing a new style of control surface with more power. However this introduced a new resonant mode, and a number of planes were lost before this was discovered.

All of these effects are often mentioned in conjunction with the term "compressibility", but in a manner of speaking, they are incorrectly used. From a strictly aerodynamic point of view, the term should refer only to those side-effects arising as a result of the changes in airflow from an incompressible fluid (similar in effect to water) to a compressible fluid (acting as a gas) as the speed of sound is approached. There are two effects in particular, wave drag and critical mach.

Wave drag is a sudden rise in drag on the aircraft, caused by air building up in front of it. At lower speeds this air has time to "get out of the way", guided by the air in front of it that is in contact with the aircraft. But at the speed of sound this can no longer happen, and the air which was previously following the streamline around the aircraft now hits it directly. The amount of power needed to overcome this effect is considerable. The critical mach is the speed at which some of the air passing over the aircraft's wing becomes supersonic.

At the speed of sound the way that lift is generated changes dramatically, from being dominated by Bernoulli's principle to forces generated by shock waves. Since the air on the top of the wing is traveling faster than on the bottom, due to Bernoulli effect, at speeds close to the speed of sound the air on the top of the wing will be accelerated to supersonic. When this happens the distribution of lift changes dramatically, typically causing a powerful nose-down trim. Since the aircraft normally approached these speeds only in a dive, pilots would report the aircraft attempting to nose over into the ground.

All of these effects have adverse effects on the control or performance of the plane. For this reason it's common to see references to aircraft that suffer from compressibility. The P-38 and Zero are particularly common examples, although in fact they are both bad ones.

ABSOLUTE AND KINEMATIC VISCOSITY

There are two ways to measure the viscosity of a fluid. One is to rotate a cup ordisk in the fluid and measure the rotational speed along with the torque required tokeep it rotating.

The density of the fluid doesn't play any part in the measurement.Viscosity is the torque divided by the speed and divided bythe contacting surfacearea. When measured this way, the viscosity is called "absolute viscosity." Itsunits are centipoise, pascal-seconds, or lb/ft-sec.

Another way to measure viscosity is to allow a certain volume of fluid to drain by gravity out of a container through a capillary tube or some other restriction.The time to drain is directly related to viscosity and is often reported in "seconds."

CONTINUITY EQUATION

Even though a detailed analysis of fluid flow can be extremely difficult, the basic concepts involved in fluid flow problems are fairly straightforward. These basic concepts can be applied in solving fluid flow problems through the use of simplifying assumptions and average values, where appropriate.

Even though this type of analysis would not be sufficient in the engineering design of systems, it is very useful in understanding the operation of systems and predicting the approximate response of fluid systems to changes in operating parameters. The basic principles of fluid flow include three concepts or principles; the first two of which the student has been exposed to in previous manuals. The first is the principle of momentum(leading to equations of fluid forces) which was covered in the manual on Classical Physics.

The second is the conservation of energy (leading to the First Law of Thermodynamics) which was studied in thermodynamics. The third is the conservation of mass (leading to the continuity equation) which will be explained in this module.

PROPERTIES OF FLUIDS

A *fluid* is any substance which flows because its particles are not rigidly attached to one another. This includes liquids, gases and even some materials which are normally considered solids, such as glass. Essentially, fluids are materials which have no repeating crystalline structure. Several properties of fluids were discussed in the Thermodynamics section of this text.

These included temperature, pressure, mass, specific volume and density. *Temperature* was defined as the relative measure of how hot or cold a material is. It can be used to predict the direction that heat will be transferred. *Pressure* was defined as the force per unit area. Common units for pressure are pounds force per square inch (psi).

Mass was defined as the quantity of matter contained in a body and is to be distinguished from weight, which is measured by the pull of gravity on a body. The *specific volume* of a substance is the volume per unit mass of the substance. Typical units are ft^3/lbm. *Density*, on the other hand, is the mass of a substance per unit volume.

Typical units are lbm/ft^3. Density and specific volume are the inverse of one another. Both density and specific volume are dependant on the temperature and somewhat on the pressure of the fluid. As the temperature of the fluid increases, the density decreases and the specific volume increases.

Since liquids are considered incompressible, an increase in pressure will result in no change in density or specific volume of the liquid. In actuality, liquids can be slightly compressed at high pressures, resulting in a slight increase in density and a slight decrease in specific volume of the liquid.

The properties outlines below are general properties of fluids which are of interest in engineering. The symbol usually used to represent the property is specified together with some typical values in SI units for common fluids. Values under specific conditions (temperature, pressure etc.) can be readily found in many reference books. The dimensions of each unit is also give in the MLT system (see later in the section on dimensional analysis for more details about dimensions.)

Density

The density of a substance is the quantity of matter contained in a unit volume of the substance. It can be expressed in three different ways.

Mass Density

Mass Density, ρ, is defined as the mass of substance per unit volume.

Units: Kilograms per cubic metre, kg/m^3 (or $^{kgm-3}$)

Dimensions: ML^{-3}

Typical values: Water = 1000^{kgm-3}, Mercury = 13546^{kgm-3} Air = 1.23^{kgm-3}, Paraffin Oil = 800^{kgm-3}. (at pressure =1.013 = 1.013×10^{-5} Nm^{-2} and Temperature = 288.15 K.)

Specific Weight

Specific Weight ω, (sometimes, and sometimes known as *specific gravity*) is defined as the weight per unit volume.

or

The force exerted by gravity, g, upon a unit volume of the substance. The Relationship between g and ω can be determined by Newton's 2ndLaw, since weight per unit volume = mass per unit volume g,

$$\omega = \rho g$$

Units: Newton's per cubic metre, N/M^3 (or Nm^{-3})

Dimensions: $ML^{-2}T^{-2}$.

Typical values:

Water = 9814 Nm^{-3}, Mercury = 132943 Nm^{-3}, Air = 12.07 Nm^{-3}, Paraffin Oil =7851 Nm^{-3}

Relative Density

Relative Density, σ, is defined as the ratio of mass density of a substance to some standard mass density. For solids and liquids this standard mass density is the maximum mass density for water (which occurs at 4°c) at atmospheric pressure.

$$s = \frac{\rho_{\text{sibstamce}}}{\rho_{H_2O(\omega 4°c)}}$$

Units: None, since a ratio is a pure number.

Dimensions: 1.

Typical values: Water = 1, Mercury = 13.5, Paraffin Oil =0.8.

Viscosity

Viscosity, m, is the property of a fluid, due to cohesion and interaction between molecules, which offers resistance to sheer deformation. Different fluids deform at different rates under the same shear stress. Fluid with a high viscosity such as syrup, deforms more slowly than fluid with a low viscosity such as water. All fluids are viscous, "Newtonian Fluids" obey the linear relationship given by Newton's law of viscosity which we saw earlier,

$$\tau = \mu \frac{du}{dy},$$

where τ is the shear stress,

Units: Nm^{-2}; $kg\ m^{-1}s^{-2}$

Dimensions $ML^{-1}T^{-2}$.

$$\frac{du}{dy}$$

is the velocity gradient or rate of shear strain, and has

Units: radianss^{-1}

Dimensions t^{-1}

μ is the "coefficient of dynamic viscosity"-see below.

Coefficient of Dynamic Viscosity

The Coefficient of Dynamic Viscosity, μ, is defined as the shear force, per unit area, (or shear stress τ), required to drag one layer of fluid with unit velocity past another layer a unit distance away.

$$\mu = \tau \frac{du}{dy} = \frac{\text{Force}}{\text{Area}} \frac{\text{Velocity}}{\text{Distance}} = \frac{\text{Force} \times \text{Time}}{\text{Area}} = \frac{\text{Mass}}{\text{Length} \times \text{Area}}$$

Units: Newton seconds per square metre, Nsm^{-2} or Kilograms per meter per second, $kgm^{-1}\ s^{-1}$. (Although note that μ is often expressed in Poise, P, where 10 P = 1 $kgm^{-1}\ s^{-1}$.)

Typical values: Water $= 1.14 \times 10^{-3}$ Air $= 1.78 \times 10^{-5}\ kgm^{-1}s^{-1}$, Mercury = 1.552 $kgm^{-1}\ s^{-1}$,

Paraffin Oil = 1.9 $kgm^{-1}\ s^{-1}$.

Kinematic Viscosity

Kinematic Viscosity, ν, is defined as the ratio of dynamic viscosity to mass density.

$$\nu \frac{\mu}{\rho}$$

Units: square metres per second, m^2s^{-1} (Although note that n is often expressed in Stokes, St, where,

$$10^4 \text{ St} = 1\,m^2s^{-1}), \text{ St} = 1\,m^2s^{-1})$$

Dimensions: $L^2 T^{-1}$.

Typical values:

Water $=1.14 \times 10^{-6}\,m^2s^{-1}$, Air $= 1.46 \times 10^{-5}\,m^2s^{-1}$, Mercury $=1.145 \times 10^{-4}\,m^2s^{-1}$,

Paraffin Oil $= 2.375 \times 10^{-3}\,m^2s^{-1}$

BUOYANCY

Buoyancy is defined as the tendency of a body to float or rise when submerged in a fluid. Weall have had numerous opportunities of observing the buoyant effects of a liquid. When we goswimming, our bodies are held up almost entirely by the water. Wood, ice, and cork float onwater. When we lift a rock from a stream bed, it suddenly seems heavier on emerging from thewater. Boats rely on this buoyant force to stay afloat. The amount of this buoyant effect wasfirst computed and stated by the Greek philosopher Archimedes. When a body is placed in afluid, it is buoyed up by a force equal to the weight of the water that it displaces.

KINEMATIC QUANTITIES OF FLOW FIELDS

FLOW LINES OF A VELOCITY FIELD

The considerations carried out in for fluid elements for the com-putation of path lines and sweepings paths have to be separated strictly from considerations for the determination of the flow lines of a flow field. Although for stationary flows flow lines, path lines and sweeping paths are identical, this does not justify to neglect the fundamental differences or even worse to assume that they do not exist. Only by a clear separation of the con-siderations it becomes generally comprehensible why under the steady-state condition of a flow field the above-stated identities of flow lines, path lines and sweeping paths exist. When one considers the non-stationary flow field $U_j(x_i, t)$, flow lines can be defined for this field at any moment in time t so that space curves run parallel to the velocity vector which are defined as flow lines at each point. In the case that one considers initially a two-dimensional flow field, for which all velocity vectors lie parallel to the plane $x_1 - x_2$, then the above-indicated definition of the flow line leads to the following defining equation:

$$\frac{d(x_2)_\psi}{d(x_1)_\psi} = \frac{U_2}{U_1} \text{ for the time } t = \text{const}$$

The relation means that the gradient of the flow line is equal to the tangent of the angle formed by the velocity vector with the axis x_1 at the moment in time t. The index ψ taken up in the equation indicates that the stated coordinates $x_1 - x_2$ describe the flow line ψ.

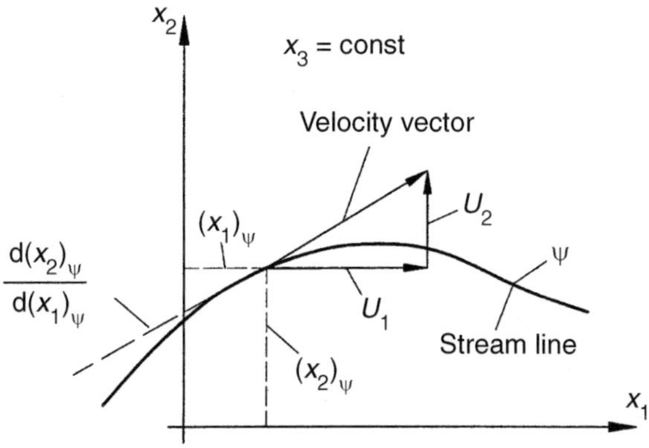

Fig. Sketch for clarifying the defining equation for the flow line of a flow field

When one considers the defining equation for the flow line of a two dimensional flow field, it becomes comprehensible that in general for each moment in time t the quotient of U_2 and U_1 is a function of x_1 and x_2. The thus resulting differential equation has to be solved in order to conserve the equation for the flow line. This will be explained on the basis of an example that starts from the following two-dimensional velocity field:

$$U_1 = x_1 (1 + 2t); \; U_2 = x_2 \text{ and } U_3 = 0$$

Introduced into the defining equation for the flow line one obtains:

$$\frac{d(x_2)_\psi}{d(x_1)_\psi} = \frac{U_2}{U_1} = \frac{x_2}{x_1(1+2t)}$$

or

$$\frac{d(x_2)_\psi}{x_2} = \frac{d(x_1)_\psi}{x_1} \frac{1}{2+2t}$$

By integration results:

$$\ln(x_2)_\psi = C + \frac{1}{1+2t} \ln(x_2)_\psi$$

Considering the flow line, passing at time $t = 0$ at the point (1, 1, 0) C = 0 results and thus:

$$(x_2)_\psi = [(x_1)_\psi]^{\frac{1}{1+2t}}$$

In the case that there is a three-dimensional velocity field, the above considerations, that were indicated for the projections of the flow lines into the planes $x_1 - x_2$., can be accepted in an analogous way. For the projections into the planes $x_1 - x_3$ – and $x_2 - x_3$. relations analogous to the defining equation result:

$$\frac{d(x_3)_\psi}{d(x_1)_\psi} = \frac{U_3}{U_1}$$

$$\frac{d(x_3)_\psi}{d(x_2)_\psi} = \frac{U_3}{U_2}$$

Thus the defining equations of the flow lines of a velocity field can be stated as follows:

$$\frac{d(x_1)_\psi}{U_1} = \frac{d(x_2)_\psi}{U_2}, \quad \frac{d(x_1)_\psi}{U_1} = \frac{d(x_3)_\psi}{U_3}, \quad \frac{d(x_2)_\psi}{U_2} = \frac{d(x_3)_\psi}{U_3}$$

or rewritten as:

$$\frac{d(x_1)_\psi}{U_1} = \frac{d(x_2)_\psi}{U_2} = \frac{d(x_3)_\psi}{U_3}$$

These differential relations for the flow line of a velocity field hold at each moment in time t. Their solution leads to a relation $(x_3)_\psi = \psi(x_1, x_2)$, which describes a space curve, the three-dimensional flow line.

Probably the most simple way to solve the law of differential equations is to seek a parameter solution $(x_1)_\psi = x_1(s)$, here s is a parameter, whose value at a certain reference point of the flow line is equal to zero and which takes increasing values along the flow line and in flow direction. When all values $-\infty < s < \infty$ are passed, through a presentation of the entire flow line is obtained.

By introducing s one obtains:

$$\frac{d(x_j)_\psi}{ds} = U_j(x_i, t) \text{ for } t = \text{const and } j = s$$

a relation which represents for each coordinate $(x_j)_\psi$ a differential equation $j = 1, 2, 3$ describing the flow lines in space for t = const. If the flow line passing through the space point $[x_0]_j$ at time t is sought, $s = 0$, results from integrating the three differential equations, when $x_j(t) = x_{j,0}$ From this results the entire flow-line field as:

$$(x_j)_\psi = \psi_j(x_{0,j}, t, s)$$

In order to demonstrate the way of proceeding in determining three-dimensional flow-line fields, the following velocity field is to be considered again:

$$U_1 = x_1(1 + t), U_2 = -x_2 \text{ and } U_3 = -x_3 t$$

Thus a law/principle/theorem of differential equations for the flow lines of this velocity field results:

$$\frac{d(x_1)_\psi}{ds} = x_1(1 + t) \quad \frac{d(x_2)_\psi}{ds} = -x_2 \quad \frac{d(x_3)_\psi}{ds} = -x_3 t$$

The integration of these equations yields:

$(x_1)_\psi = C_1 \exp[(1 + t)s]$,
$(x_2)_\psi = C_2 \exp[-s]$,
$(x_3)_\psi = C_3 \exp[-ts]$.

Searching for the flow line that passes through the point(1, 1, 1) one can chose this point as a point of reference and set $s = 0$ for $(x_i)_\psi = 1$ From this the following results for the integration constants:

$C_1 = C_2 = C_3 = 1$

One obtains thus:

$(x_1)_\psi = \exp[(1 + t)s]$,
$(x_2)_\psi = \exp[-s]$,
$(x_3)_\psi = \exp[-ts]$.

For the time $t = 0$ a flow line path results

$(x_1)_\psi = \exp[s]$, $(x_2)_\psi = \exp[-s]$, $(x_3)_\psi = 1$

So one obtains a flow line passing in the plane $x_3 = 1$ and described there by the relation

$$(x_2)_\psi = \frac{1}{(x_1)_\psi}$$

The entire flow line field is obtained when one introduces for $s = 0$ arbitrary position coordinates $(x_{0,i})$ so that for the position coordinates holds:

$(x_1)_\psi = (x_1)_{\psi,0} \exp[(1 + t)s]$,
$(x_2)_\psi = (x_2)_{\psi,0} \exp[-s]$,
$(x_3)_\psi = (x_3)_{\psi,0} \exp[-ts]$.

This set of equations indicates at each time t the flow lines passing through the point $\{x_j\}_{\psi,0}$ There the parameter s has the value $s = 0$. and for all other points of the flow line a value equal to zero.

A clear assignment is thus occurring which guarantees that flow lines never intersect, as otherwise two velocities would exist simultaneously at this point of intersection.

Thus occurring which guarantees that flow lines never intersect, as otherwise two velocities would exist simultaneously at this point of intersection. This is excluded because of the existence of clear velocity fields (except for stagnation points and singularities).

In the preceding section it was emphasized that in general flow lines, path lines and sweeping lines are not identical and the computed examples have made this clear have confirmed this.

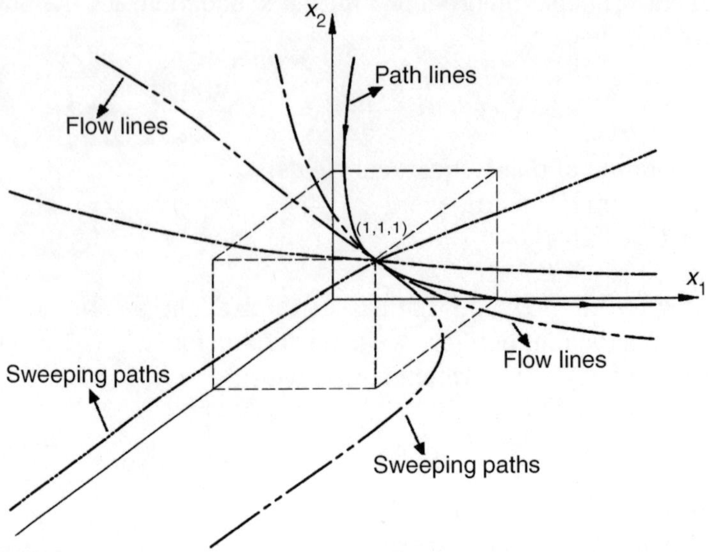

Fig. Comparison of flow lines, path lines and sweeping paths

For stationary flows all three lines are identical, *i.e.*

- For stationary flow fields marked fluid elements move along flow lines, *i.e.* flow lines are equal to the path lines
- For stationary flow fields flow lines can be made visible by locally injected tracer particles, *i.e.* flow lines are equal to the sweeping paths

For non-stationary flows generally flow lines, path lines and sweeping paths are different space curves.

Stream function and Flow Lines of two-dimensional Flow Fields

For two-dimensional incompressible velocity fields $\{U_j\} = \{U_1, U_2, 0\}$ the stream function can be introduced as a field quantity. It is defined as follows:

$$U_1 = \frac{\partial_\psi}{\partial x_2} \text{ und } U_2 = \frac{\partial_\psi}{\partial x_1}$$

and is computed from the velocity field over the following line integrals:

$$\psi - \psi_0 = \int_{x_2,0}^{x_2} U_1 \mathrm{d}x_2 \text{ for } \mathrm{x}_1 = \text{const}$$

$$\psi - \psi_0 = \int_{x_1,0}^{x_1} U_1 \mathrm{d}x_1 \text{ for } \mathrm{x}_2 = \text{const}$$

The above equations show that the difference existing between two stream-

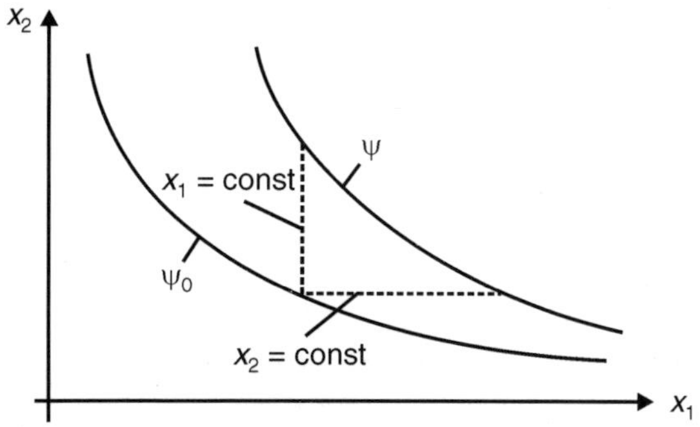

Fig. Explanation of the defining equations of stream function

function values is a measure for the volume flow that runs between two lines each with a constant stream-function value each, where a depth of the linear measure 1 is considered vertically to the plane $x_1 - x_2$. Accordingly the integrations along the line drawn and stated in the equations x_1 = const and x_2 = const yield identical volume-flow values, when the upper integration limits x_1 and x_2 are chosen such that the difference $\psi - \psi_0$ is equally large in both integrals.

For a fluid with ρ = const, *i.e.* for an ideal fluid, the identity of the integrals represents a mass-conservation relation.

When one carries out the integrations stated in the equations along a line ψ = const *i.e.* for $d\psi = 0$, one obtains:

$$U_1 d(x_2)_\psi = U_2 d(x_1)\psi$$

or transcribed:

$$\frac{d(x_1)_\psi}{U_1} = \frac{d(x_2)_\psi}{U_2}$$

From the above indicated relation it can be said that the stream function by a comparison with the defination equation for the flow line stated in equation that the stream function defined for two-dimensional flow fields yields flow lines ψ for ψ = const.

While for the kinematic considerations is arbitrary mathematics set-ups for the velocity field could be used, the intro-duction of the stream function requires a reduction of the considerations to those velocity fields which have to fulfill physical conditions. Physically possible flow fields have to fulfill the mass-conservation law which can be formulated for ideal fluids (ρ = const) as follows:

$$\frac{\partial U_i}{\partial x_i} = \frac{\partial U_1}{\partial x_1} + \frac{\partial U_2}{\partial x_2} + \frac{\partial U_3}{\partial x_3} = 0$$

That a stream function fulfills the mass-conservation law for two-dimensional flows automatically, can easily be checked by inserting the definition in equation where the law/principle of Schwarz has to be applied.

When a flow field does not fulfill the mass-conservation law, the integrations to be carried out according to the equations result in solutions contradicting one another. This can be shown for the employed two-dimensional flow field which does not fulfill the mass-conservation law (*i.e.* the continuity equation):

$$U_1 = x_1;\ U_2 = x_2(1 + 2t) \text{ and } U_3 = 0$$

When one carries out the integration stated in equation in spite of this, from the defining equations for the stream function, *i.e.* from the relations results:

$$\frac{\partial_\psi}{\partial x_2} = x_1 \text{ and } \frac{\partial_\psi}{\partial x_1} = -x_2(1+2t)$$

By integration of these equations one obtains for the stream function:

$$\psi = x_1 x_2 + F(x_1, t)\ \psi = -x_1x_2(1 + 2t) + G(x_2, t)$$

or expressed otherwise:

$x_1x_2 + F(x_1, t) \neq -x_1 x_2(1 + 2t) + G(x_2, t)$

or $2(t + 1)x_1 x_2 \neq G(x_2, t) - F(x_1, t)$

A comparison of the results of both the integrations, which yield the equations shows the contradiction resulting for the stream function.

The time dependence showing on the right side of the relation is missing on the left side. This results from the fact that the velocity field indicated according to although mathematically clearly defined, cannot exist physically; the velocity field does not fulfill the requirements determined by the mass-conservation law for ρ = const When one considers on the other hand the velocity field:

$U_1 = \exp[x_1(1 + t)];\ U_2 = -x_2(1 + t)\exp[x1\ (1 + t)];\ \ U_3 = 0$ for which the relation is fulfilled, where it holds:

$$\frac{\partial U_1}{\partial x_1} = (1 + t)\exp[x_1(1 + t)] \text{ and}$$

$$\frac{\partial U_2}{\partial x_2} = -(1 + t)\exp[x_1(1 + t)]$$

one obtains from the defining equations for the stream function :

$$\frac{\partial \psi}{\partial x_2} = \exp[x_1(1 + t)] \text{ and}$$

$$\frac{\partial \psi}{\partial x_1} = x_2(1 + t)\exp[x_1(1 + t)]$$

the following solution for the stream function ψ:

$$\psi = x_2 \exp[x_1(1 + t)] + C$$

If one knows the value of the stream function for $x_{1,0}$ and $x_{2,0}$, it holds

$$\psi_0 = x_2, 0 \exp[x_1, 0 (1 + t)] + C$$

$$\psi - \psi_0 = x_2 \exp[x_1(1 + t)] - x_{2,0} \exp[x_{1,0} (1 + t)]$$

The obtained result can also be computed for $x_1 = x_{1,0}$ = const . from the relation:

$$\psi - \psi_0 = (x_2 - x_{2,0}) \exp[x_{1,0}(1 + t)] \text{ for } x_1 = x_{1,0}$$

This is indicated by the distribution $\psi(x_2)$ at the location $x_{1,0}$ an.

When one wants to determine the flow line path in the plane $x_1 - x_2$ – one has to consider y as a parameter in equation and derive the relation $x_2 - x_1$ – for this parameter from equation. Here $\psi_0, x_1,0$ and $x_2, 0$ are constants that can be chosen freely.

Divergence of a Flow Field

In this section mathematics operators shall be explained that are known from vector analysis and that can be applied to flow fields. Their derivations will also be repeated and considered as concerns their physical meaning. Divergence of the stream field $\vec{U}$:

$$\frac{\partial U_i}{\partial x_i} = \frac{\partial U_1}{\partial x_1} + \frac{\partial U_2}{\partial x_2} + \frac{\partial U_3}{\partial x_3}$$

The above defined divergence of a flow field is to be considered as a scalar field which, in the presence of a steady velocity field, is defined at each point in space and can be computed from the velocity field that can be assumed to be known.

When one wants to perceive the physical meaning of the operator U_i applied to the velocity field $\frac{\partial}{\partial x_i}$ x_i the consideration of a fluid element, as stated in Fig.4.9, is recommended.

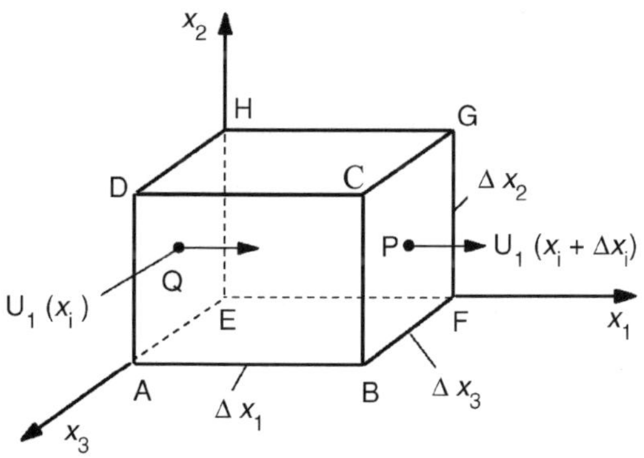

Fig. Fluid element for explaining the physical meaning of $\frac{\partial U_i}{\partial x_i}$

The edge lengths Δx_i were assumed to be very small, so that a velocity vector can be assigned to each surface such that this vector indicates with which velocity the considered surface of a fluid element moves. Accordingly, the surface AEHD moves in the direction x_1 with the velocity component of the velocity field present in point Q *i.e.* with $U_1(x_i) = U_1(x_1, x_2, x_3)$. In comparison, the surface BFGC moves with the velocity component $U_1(x_i + Dx_i, x_2, x_3)$.present in point P. This velocity component can be expressed by a Taylor series expansion as follows:

$$U_1(x_1 + Dx_1, x_2, x_3) = U_1(x_1) + \left(\frac{\partial U_1}{\partial x_1}\right)\Delta x_1 + \frac{1}{2}\left(\frac{\partial^2 U_1}{\partial x_1^2}\right)\Delta x_1^2 + \ldots$$

The difference velocity between the surfaces AEHD and BFGC can thus be computed as:

$$\Delta U_1(x_i, \Delta x_i) = \left(\frac{\partial U_1}{\partial x_1}\right)\Delta x_1 + \frac{1}{2}\left(\frac{\partial^2 U_1}{\partial x_1^2}\right)\Delta x_1^2 + \ldots$$

As a consequence of this velocity difference a volume increase or a volume decrease results, depending on which signs the derivations of the velocity fields, which can be stated as follows in a first approximation by multiplication with the surface $\Delta x_2\,\Delta x_3$, neglecting the terms of second and higher order in Δx_1:

$$\frac{d}{dt}(\delta V_1)_{\Re} = \Delta U_1(x_i)(\Delta x_2 \Delta x_3) = \frac{\partial U_1}{\partial x_1}\Delta x_1(\Delta x_2 \Delta x_3)$$

On the basis of simultaneously existing gradients of the velocity field in the directions x_2 – and x_3 in addition volume changes occur in the time unit, which again can be stated in a first approximation as follows:

$$\frac{d}{dt}(\delta V_2)_{\Re} = \frac{\partial U_2}{\partial x_2}\Delta x_2(\Delta x_1 \Delta x_3) \text{ and}$$

$$\frac{d}{dt}(\delta V_3)_{\Re} = \frac{\partial U_3}{\partial x_3}\Delta x_3(\Delta x_1 \Delta x_2)$$

so that the entire volume change that can be expected in a flow field for a fluid element in the time unit can be indicated as follows:

$$\frac{d}{dt}(\delta V_3)_{\Re} = \sum_{\alpha=1}^{3}\frac{d}{dt}(\delta V_\alpha)_{\Re} = \frac{\partial U_i}{\partial x_i}(\delta V_{\Re})$$

or can be written as:

$$\frac{\partial U_i}{\partial x_i} = \frac{1}{\delta V_{\Re}}\frac{d}{dt}(\delta V_{\Re})$$

This relation makes emphasizes the physical significance of the divergence of a velocity field [3]. In accordance the divergence of a velocity field states how big the volume change of a fluid element is that occurs per time and volume unit at a certain position in a flow field.

At such locations of the flow field where the divergence of a velocity vector is equal to zero, there is no temporal volume change locally for a fluid element moving in the velocity field. When the divergence in sub-domains of the velocity field is computed negatively, a fluid element experiences volume decreases in these domains.

When one carries out considerations concerning the physical significance of the divergence of a velocity field at a stationary volume element of a fluid, inflows and outflows occur through the surfaces of the considered volume because of the existing velocity field, to an extent that the volume flowing in per time unit can be stated:

$$\dot{V}_{\text{inflow}} = U_i \, \Delta x_j \, \Delta_{xk} \; i = j, k$$

For the volume flowing out it can be computed:

$$\dot{V}_{\text{Out flow}} = \left[U_i + \frac{\partial U_i}{\partial x_i} \right] \Delta x_j \Delta x_k \;\; i = j, k$$

The relation makes clear that the sumation for $i = 1$ to 3 is stated in a sufficient6ly comprehensible way by the double index in $\partial U_i / \partial x_i$

The difference of inflows and outflows, considering DV = Dx_1, Dx_2 , Dx_3 can be computed as:

$$\Delta \dot{V} = \dot{V}_{\text{In flow}} - \dot{V}_{\text{Out flow}} = -\frac{\partial U_i}{\partial x_i} \Delta V$$

This relation makes clear that the presence of a positive divergence of the velocity field in volume is equal to a source, as more "'fluid volume"' is flowing out than flowing in. When, however, the divergence of a velocity field is negative, a sink occurs, as then the inflow in "'volume"' has to be larger than the outflow.

BOUNDARY CONDITION OF KINEMATICS

At an impermeable boundary, the flow of fluid *relative to the boundary* must be tangential to it. If the boundary is fixed in space, this means that the component of fluid velocity normal to the boundary must be zero. However, if the boundary is moving, then the normal component of the fluid must be equal to the velocity of the boundary normal to itself. This applies whether there is a solid on one side of the boundary or not. If not, then we have the case of a free surface of the fluid, such as the sea surface.

Let $\vec{U}$ represent the velocity of the fluid boundary and $\vec{V}$ be the fluid velocity of the boundary. Then the kinematic boundary condition is simply,

$$\left(\vec{U}-\vec{V}\right)\cdot\vec{n}=0$$

where $\vec{n}$ is the unit normal to the boundary.

The kinematic boundary condition for an impermeable boundary only, *i.e.* a boundary *across* which there is no flux of matter of the macroscopic scale. In those cases where a change of state of the fluid occurs at the boundary or where the boundary surface displays a selective permeability phenomenon, the condition is no longer rigourously applicable.

A specific case in point is the effect of a flux of evaporation at the sea surface. This is certainly important in thermodynamic problems although in nearly all kinematic or dynamical problems the effect of evaporation can be ignored so far as the kinematic boundary condition is concerned.

The boundary condition can be written in a more convenient form for a free surface of the fluid. In any physically real situation a free surface simply represents the boundary between two different fluids, *e.g.* water and air, water and steam, oil and water, etc.

That is, the free surface exists by virtue of a difference in the neighboring fluids. The difference can be a chemical difference or merely a physical difference (difference in phase of the same substance, *e.g.* water/water vapor). The important thing in connection with our present discussion is that there exist certain physical or thermal properties of the two fluids which are continuous across the free surface. An example is temperature.

Pressure is another example in those cases where the effect of surface tension can be ignored.

Suppose $f(x,y,z,t)$ represents the field of some property in the fluid concerned and $f'(x,y,z,t)$ the field of the same property in the adjacent fluid. If the property f is one which is continuous across the free surface (which delineates the boundary between these two fluids) then,

$$f(x,y,z,t) = f'(x,y,z,t)$$

it at the boundary. If the functions f and f' are known, then Eq.actually defines the boundary, *i.e.*

$$F'(x,y,z,t) = \mathrm{f} - \mathrm{f}' = 0$$

is the implicit equation of the free surface. Usually this is written in the explicit form

$$z=\eta\left(x,y,t\right),$$

where η denotes the function representing the elevation of the free surface above some reference plane at position x, y and time t.

The time rate of change of the property f following the motion of the fluid at the boundary is $\left(Df/Dt\right)_z=\eta$. The condition that the fluid at the boundary flows along the boundary and never leaves the boundary is equivalent to the condition that the individual rate of change of f at the boundary is equivalent to the rate of change of *f*′ on the other side of the boundary as one follows with the particle motion of the first fluid. Thus,

$$\frac{Df}{Dt} = \left(\frac{Df}{Dt} - \frac{Df'}{Dt}\right)_{z=n} = 0$$

represents the kinematic boundary condition. In view of Eq., it will be noted that

$$F = \eta(x, y, t) - z$$

so that Eq. can be written as,

$$\frac{\partial \eta}{\partial t} + u_\eta \frac{\partial \eta}{\partial x} + v_\eta \frac{\partial \eta}{\partial y} - w_\eta = 0$$

or,

$$w_\eta = \frac{\partial \eta}{\partial t} + u_\eta \frac{\partial \eta}{\partial x} + v_\eta \frac{\partial \eta}{\partial y}$$

where u_η, v_η, w apply at the boundary.

If the free surface is *stationary* (*i.e.* $\partial \eta / \partial t = 0$) and slopes only in the x-direction, then,

$$w_\eta = u_\eta \frac{\partial \eta}{\partial x}.$$

This expression can be verified by reference to Fig. below. If a particle starting at position A is to remain on the surface as itmoves over the differential distance *ds*, then we must have

$$w_\eta dt = \frac{\partial \eta}{\partial x} u_\eta dt$$

which is equivalent.

If the surface has no slope but is ascending at the rate of $\partial \eta / \partial t$ then it is obvious that in order that the fluid remain at the surface, then w_η must satisfy the relation

$$w_\eta = \frac{\partial \eta}{\partial t}.$$

Equation is applicable when both effects are present, including the effect of surface slope in the y-direction as well as the x-direction.

In many problems the non-linear terms in equation are of second order compared with $\partial \eta / \partial t$ and equation represents a good approximation. However, one should always investigate the order of magnitude of these terms before making such an approximation.

APPLICATION OF THE BOUNDARY CONDITION FOR LAPLACIAN FLOW ADJACENT TO A SOLID CYLINDER

Consider a solid cylinder of circular cross-section which travels at constant velocity *U* normal to its axis through a fluid of very vast extent. That is, it is presumed that the linear dimensions of the fluid are many orders of magnitude greater than the diameter of the cylinder. The flow of fluid induced by the motion of the cylinder can be expected to be essentially two-dimensional, considering that the cylinder is very long compared with its diameter. This implies that the picturre of flow is the same in any plane which is normal to the axis of the cylinder.

If the flow is presumed to be Laplacian (*i.e.* divergenceless and irrotational) then we can evaluate the pattern of flow by the methods outlined in Art. 1.11, making use of the appropriate boundary conditions. It is sufficient to state at this point that Laplacian flow adjacent to the cylinder is a rather idealized type of motion, however it does possess certain characteristics which warrant its investigation. Practically speaking, the flow would exist immediately after an impulse is imparted to a cylinder which was originally at rest in a static fluid. The boundary condition to be satisfied is equation. We will take the origin of the reference frame at the center of the cylinder, and thus refer all quantities relative to the moving cylinder. The normal component of the fluid velocity to the surface is,

$$\left(V_r\right)_{r=a} = U\cos\theta.$$

Since the flow is presumed to be Laplacian, it follows that V_r can be expressed as,

$$\left(V_r\right) = -\frac{\partial\phi}{\partial_r}.$$

Consequently, equation can be written as,

$$\left(\frac{\partial\phi}{\partial r}\right)_{r=a} = -U\cos\theta$$

The transform $W = f(Z)$, which is capable of satisfying condition, is,

$$W = AZ^n, \qquad A = \text{const.}$$

which implies that,

$$\phi = Ar^n \cos n\theta$$

$$\phi = Ar^n \sin n\theta.$$

Applying condition (234a) gives,

$$An\, a^{(n-1)} \cos n\theta = -U \cos\theta$$

which can be satisfied only if,

$$n = \pm 1$$

$$A = -\frac{U}{na^{n-1}}.$$

The velocity at distance *r* is,

$$V = \sqrt{\left(\frac{\partial \phi}{\partial r}\right)^2 + \left(\frac{1}{r}\frac{\partial \phi}{\partial \theta}\right)^2} = Anr^{n-1}$$

or,

$$V = U\left(\frac{r}{a}\right)^{n-1}.$$

In order to select the proper sign of *n* in equation we introduce a second condition which is simply that the motion is to be confined to the neighborhood of the cylinder; that is, the fluid is not significantly disturbed at a great distance from the cylinder. This requires that we take n = –1 and, consequently,

$$\phi = U\frac{a^2}{r}\cos\theta,$$

$$\psi = -U\frac{a^2}{r}\sin\theta,$$

$$V = U\left(\frac{a}{r}\right)^2.$$

The streamlines and isolines of velocity potential repesented by equations. The families of isolines are circles which pass through the point 0 if extended inside the cylinder.

It is of interest to compute the kinetic energy of the fluid motion induced by the passage of the cylinder through the fluid. This is given by the integral,

$$E_k = \frac{1}{2}\rho\int_a^{\infty}\int_0^{2\pi} V^2\, r\, dr\, d\theta,$$

where ρ is the mass density of the fluid, presumed constant, and E_k is the total kinetic energy of that fluid contained between two planes of unit distance apart and normal to the axis of the cylinder and extending indefinitely in the *x* and *y* directions. If we carry out the integration, making use of equation, we find that

$$E_k = \frac{1}{2}\pi a^2 \rho u^2$$

or,

$$E_k = \frac{1}{2}M'U^2$$

where $M' = \rho\pi a^2$ represents the mass of fluid displaced by the cylinder per unit length of cylinder.

If the cylinder is accelerated, then a force commensurate with the mass of the cylinder *plus* the mass *M'* is required to produce the given acceleration. That is, of the kinetic energy being supplied to the cylinder, some is used to accelerate the cylinder and some is used to accelerate the water in the neighborhood of the cylinder. The effective mass of the cylinder in the fluid therefore includes the mass *M'*.

KELVIN'S THEOREM OF MINIMUM KINETIC ENERGY

Kelvin has established the remarkable theorem that Laplacian flow has less kinetic energy than any other motion consistent with the same normal motion of the boundaries of the fluid. A practical criterion can be immediately inferred from this theorem in connection with the foregoing example the effective added mass, M_1, is a minimum in the case of Laplacian flow around a cylinder. In the more usual case where turbulent flow exists in the neighborhood of the cylinder, we should expected to find that,

$$C_M \equiv \frac{M_\theta}{M} > 1$$

for a circular cylinder.

ADDITION OF VELOCITY POTENTIALS

If two velocity fields $\vec{V}_1$ and $\vec{V}_2$ can be represented by means of the velocity potentials ϕ_1 and ϕ_2, respectively, then their superposition can be represented by

$$\vec{V}_1 + \vec{V}_2 = -\nabla(\phi_1 + \phi_2).$$

In other words, the algebraic sum $(\phi_1 + \phi_2)$ defines the superimposed fields of $\vec{V}_1$ and $\vec{V}_2$. As an illustration, suppose that it is desired to determine the irrotational and divergenceless flow of fluid past a circular cylinder which is held fixed. Since the flow is presumed to be Laplacian, the pattern can be established merely by superimposing a velocity *-U* on the pattern of flow. This will effectively stop the cylinder, and at the same time alter the pattern of the streamlines and velocity potential. The velocity *-U* can be represented by the potential

$$\phi_2 = U_x = U_r \cos\theta.$$

Thus if we let,

$$\phi_1 = U\frac{a^2}{r}\cos\theta$$

from equation then the potential applicable to flow past a stationary cylinder is,

$$\phi = \phi_1 + \phi_2 = U\left(r + \frac{a^2}{r}\right)\cos\theta.$$

We could also take,

$$\psi_1 = -U\frac{a^2}{r}\sin\theta,$$
$$\psi_2 = U_y = U_r \sin\theta$$

and the superposition of these streamfunctions gives,

$$\psi = \psi_1 + \psi_2 = U\left(r - \frac{a^2}{r}\right)\sin\theta.$$

Continuityequation

The continuity equation is simply a mathematical expression of the principle of conservation of mass. For a control volume that has a single inlet and a single outlet, the principle of conservation of mass states that, for steady-state flow, the mass flow rate into the volume must equal the mass flow rate out. The continuity equation for this situation is expressed by Equation.

$$\dot{m}_{inlet} = \dot{m}_{outlet}$$
$$(\rho A v)_{inlet} = (\rho A v)_{outlet}$$

For a control volume with multiple inlets and outlets, the principle of conservation of mass requires that the sum of the mass flow rates into the control volume equal the sum of the mass flow rates out of the control volume. The continuity equation for this more general situation is expressed by Eq.

$$\sum \dot{m}_{inlet} = \sum \dot{m}_{outlet}$$

One of the simplest applications of the continuity equation is determining the change in fluid velocity due to an expansion or contraction in the diameter of a pipe.

Example: Continuity Equation - Piping Expansion

Steady-state flow exists in a pipe that undergoes a gradual expansion from a diameter of 6 in. to a diameter of 8 in. The density of the fluid in the pipe is constant at 60.8 lbm/ft3. If the flow velocity is 22.4 ft/sec in the 6 in. section, what is the flow velocity in the 8 in. section?

Solution: From the continuity equation we know that the mass flow rate in the 6 in. section must equal the mass flow rate in the 8 in. section. Letting the subscript 1 represent the 6 in. section and 2 represent the 8 in. section we have the following.

$$\dot{m}_1 = \dot{m}_2$$

$$\rho_1 A_1 v_1 = \rho_2 A_2 v_2$$

$$v_2 = v_1 \frac{\rho_1}{\rho_2} \frac{A_1}{A_2}$$

$$= V_t \frac{\pi r_1^2}{\pi r_2^2}$$

$$= \left(22.4 \frac{ft}{sec}\right) \frac{(3\,in)^2}{(4\,in)^2}$$

$$v_2 = 12.6 \frac{ft}{sec}$$

So by using the continuity equation, we find that the increase in pipe diameter from 6 to 8 inches caused a decrease in flow velocity from 22.4 to 12.6 ft/sec.

The continuity equation can also be used to show that a decrease in pipe diameter will cause an increase in flow velocity.

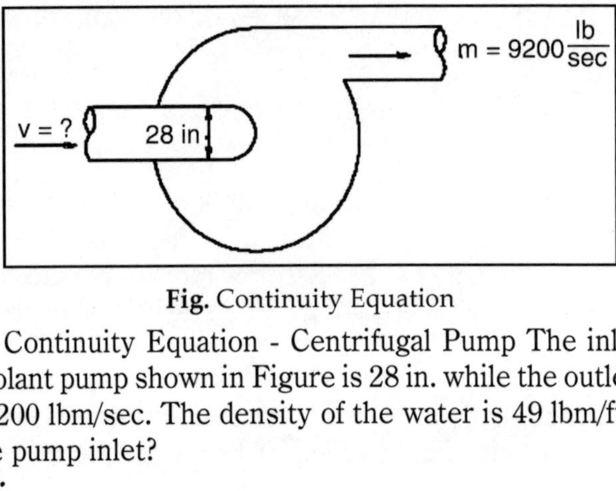

Fig. Continuity Equation

Example: Continuity Equation - Centrifugal Pump The inlet diameter of the reactor coolant pump shown in Figure is 28 in. while the outlet flow through the pump is 9200 lbm/sec. The density of the water is 49 lbm/ft3. What is the velocity at the pump inlet?

Solution:

$$A_{inlet} = \pi r^2 = (3.14)\left(14\ in\left(\frac{1 ft}{12 in}\right)\right)$$

$$= 4.28\,ft^2$$

$$\dot{m}_{inlet} = \dot{m}_{outlet} = 9200 \frac{lbm}{sec}$$

$$(\rho A v)_{inlet} = 9200 \frac{lbm}{sec}$$

$$v_{inlet} = \frac{9200\frac{lbm}{sec}}{Ap}$$

$$= \frac{9200\frac{lbm}{sec}}{(4.28\,ft^2)\left(49\frac{lbm}{ft^2}\right)}$$

$$v_{inlet} = 43.9\frac{ft}{sec}$$

The above example indicates that the flow rate into the system is the same as that out of the system. The same concept is true even though more than one flow path may enter or leave the system at the same time.

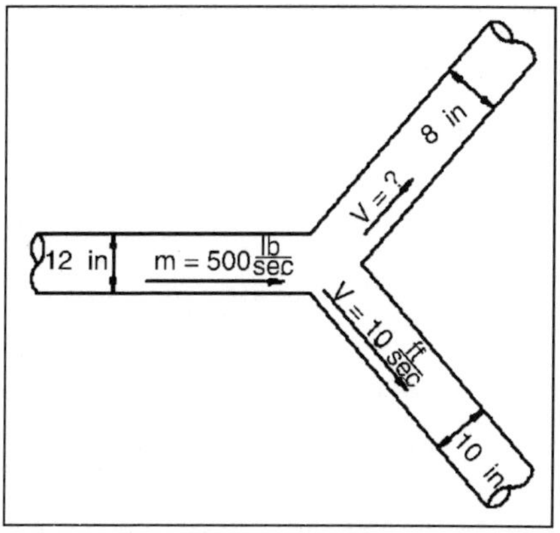

Fig. "Y" Continuity for Example Problem

The mass balance is simply adjusted to state that the sum of all flows entering the system is equal to the sum of all the flows leaving the system if steady-state conditions exist. An example of this physical case is included in the following example.

Example: Continuity Equation - Multiple Outlets. A piping system has a "Y" configuration for separating the flow as shown in Figure. The diameter of the inlet leg is 12 in., and the diameters of the outlet legs are 8 and 10 in. The velocity in the 10 in. leg is 10 ft/sec. The flow through the main portion is 500 lbm/sec. The density of water is 62.4 lbm/ft3. What is the velocity out of the 8 in. pipe section?

Solution:

$$A_8 = \pi\left(4\text{ in.}\frac{1\,ft}{12\,in.}\right)^2 - 0.349\ ft^2$$

$$A_{10} = \pi\left(5\text{ in.}\frac{1\,ft}{12\,in.}\right)^2 - 0.545\ ft^2$$

$$\sum \dot{m}_{inlets} = \sum \dot{m}_{outlets}$$

$$\dot{m}_{12} = \dot{m}_{10} + \dot{m}_8$$

$$\dot{m}_8 = \dot{m}_{12} - \dot{m}_{10}$$

$$(\rho A v)_8 = \dot{m}_{12} - (\rho A v)_{10}$$

$$v_8 = \frac{\dot{m}_{12} - (\rho A v)_{10}}{(\rho A)_8}$$

$$= \frac{500\frac{lbm}{sec}\left(62.4\frac{lbm}{ft^2}\right)(0.545\,ft^2)\left(10\frac{ft}{sec}\right)}{\left(62.4\frac{lbm}{ft^2}\right)(0.349\,ft^2)}$$

$$v_8 = 7.3\,\frac{ft}{sec}$$

KINEMATIC CLASSIFICATION OF FLUID MOTION

One way of classifying different types of motion is according to their kinematic properties. For example, there is a large class of motions dealing with divergenceless fluids (commonly referred to as motion of *incompressible fluids*. Another class is motion wherein the vorticity is everywhere zero. This is referred to as *irrotational motion*. Motions of solid bodies, representing perhaps the most widely studied class in applied mechanics, falls under the heading of *deformationless motion*.

STREAMFUNCTION FOR TWO-DIMENSIONAL FLOW

If a particular two-dimensional motion is divergenceless, implying that,

$$\frac{\partial u}{\partial x} + \frac{\partial v}{\partial y} = 0,$$

then the velocity field can be represented in terms of a single scalar function commonly referred to as the streamfunction. If we let ψ represent the streamfunction, then

$$u = -\frac{\partial \psi}{\partial y}$$

and,

$$v = -\frac{\partial \psi}{\partial y}$$

Equation is a necessary condition that such a function as implied by Eq. exist. It is readily verified that equations satisfy the condition, provided that

$$\frac{\partial}{\partial x}\frac{\partial \psi}{\partial y} = \text{ppxy}\left(\frac{\partial \psi}{\partial x}\right),$$

which holds as long as ψ possesses first and second derivatives.

VELOCITY POTENTIAL

If a particular fluid motion is irrotational, meaning that ,

$$\nabla \times \vec{V} = 0,$$

it is then possible to represent the field of flow in terms of a single scalar function ϕ as,

$$\vec{V} = -\nabla \phi$$

or, componentwise,

$$u = -\frac{\partial \phi}{\partial x},$$

$$v = -\frac{\partial \phi}{\partial y}, \text{and}$$

$$w = -\frac{\partial \phi}{\partial z}.$$

The quantity ϕ, referred to as the velocity potential, is a function of x,y,z,t in the most general case. Unlike the streamfunction, which is applicable for divergenceless, two-dimensional flow, the velocity potential may exist even for three-dimensional flow. However, the necessary condition for the existence of ϕ as defined by Eq. is somewhat more restrictive since, in effect, Eq. requires that

$$\frac{\partial w}{\partial y} = \frac{\partial v}{\partial z}, \frac{\partial u}{\partial z} = \frac{\partial w}{\partial x}, \frac{\partial v}{\partial x} = \frac{\partial u}{\partial y}.$$

LAPLACIAN FLOW

A two-dimensional field of flow which is both divergenceless and irrotational can be characterized by either one of the scalar functions ψ or ϕ. This type of flow, which bears the name of Laplace, possesses a number of interesting features which have been extensively investigated, and forms the bulk of much of the classical literature on the subject of hydrodynamics. Only some of the properties of this type of flow will be dealt with here. For a more exhaustive treatment, the reader is referred to the incomparable work of Lamb.

CONFORMAL MAPPING

It is possible at once to write down the most general solution for ψ and ϕ as governed by the equations of Art.. This can be done by introducing the concept of functions of a complex variable. If we let,

$$W = \phi + i\psi$$

and,

$$Z = x + iy$$

then,

$$W = f(Z)$$

represents the most general solution for ϕ and ψ, where f is any differentiable function at all. In any particular case, the problem is reduced to that of finding a function f which will satisfy certain boundary conditions.

The proof that Eq. satisfies the basic conditions on ϕ and ψ expressed by Eqs. is straightforward. If we differentiate W with respect to x we obtain,

$$\frac{\partial \phi}{\partial x} + i\frac{\partial \psi}{\partial x} = \frac{df}{dz}\frac{\partial z}{\partial x} = \frac{df}{dz}.$$

Differentiating W *with respect to* y, *on the other hand, gives,*

$$\frac{\partial \phi}{\partial y} + i\frac{\partial \psi}{\partial y} = \frac{df}{dz}\frac{\partial z}{\partial y} = i\frac{df}{dz}.$$

Eliminating df/dZ *between (214) and (215) yields,*

$$\frac{\partial \phi}{\partial y} + i\frac{\partial \psi}{\partial y} = i\frac{\partial \phi}{\partial x} - \frac{\partial \psi}{\partial x}.$$

This equation, however, can only be satisfied if,

$$\frac{\partial \phi}{\partial y} = -\frac{\partial \psi}{\partial x},$$

and ,

$$\frac{\partial \psi}{\partial y} = -\frac{\partial \phi}{\partial x},$$

where it is considered that the quantities ϕ and ψ are real. These equations are in accord with euations, as can be seen by equating the two expressions for u and the two expressions for v. The conditions in fact represent the Cauchy-Riemann conditions which are required if the function f(Z) is to possess a unique derivative.

The relation between W and Z stipulated by implies that for each point x, y in the Z-plane there corresponds at least one value of ϕ and ψ in the W-plane. One can map out the isolines of ϕ and ψ in the x, y plane or isolines of x and y in the ϕ,ψ plane. In either case the sets of isolines are orthogonal.

Consider an arbitrary curve C on the Z-plane. This can be mapped on the W-plane through a given transform w = F(z). Suppose the curve C and its mapped counterpart. Any point on the curve in the Z-plane can be represented by x, y or by the polar coordinates r, ϑ. The complex quantity Z can be written in terms of r and ϑ as,

$$Z = x + iy = r\cos\vartheta + i\left(r\sin\vartheta\right) = re^{i\vartheta}.$$

The corresponding point ϕ, ψ in the W-plane can be expressed by the polar coordinates $R\theta$. In terms of these quantities the complex potential is,

$$W = Re^{i\theta}$$

The solution of a problem of Laplacian flow in the vicinity of a fixed boundary is tantamount to finding that transform w = F(z) which maps the boundary as the straight line ψ = constant in the W-plane. That is, for a fixed boundary the flow must be along the boundary and consequently the latter must be a surface of constant ψ. Consider the problem of ascertaining the two-dimensional, irrotational, divergenceless flow in the neighborhood of the solid boundary formed by two planes intersecting at an arbitrary angle α.

The boundary, c – c', can be mapped on the W-plane as a straight line by use of the following transform:

$$W = z^N.$$

If we introduce the complex polar form of W and Z given by Eqs., we have from Eq.

$$Re^{i\theta} = r^n\, e^{in\vartheta},$$

which implies that,

$$R = z^N.$$

and,

$$\theta = n\vartheta$$

If we then choose the value of n *as,*

$$n = \frac{\pi}{\text{alpha}}$$

then the curve c - c' will represent the axis of ϕ on the W plane, *i.e.* the line $\phi = 0$, which satisfies the condition that the boundary is a streamline.

The streamlines for other positions in the fluid can now be mapped from the W-plane back onto the Z-plane by use of the relations. As a specific example suppose $a = 2\pi / 3$ radians, then,

$$n = 3/2$$

and,

$$r = R^{2/3},$$

$$\vartheta = 2/3\theta.$$

For any particular isoline ψ = constant, the value of R consistent with a selected ϕ can be found from the isoline of ψ in the W-plane and mapped on the Z-plane using the transformed coordinates r, ϑ from equations. The isolines of ϕ can be mapped on the Z-plane in a similar manner. The resulting orthogonal fields of ψ and ϕ. Either set of isolines determines the flow uniquely in the neighborhood of the boundary. The isolines of ψ obtained in this illustration represent a family of hyperbolas whose major axis is the isoline $\phi = 0$ and whose asymptotes are the boundary lines c – c'. If we express W as the complex potential $\phi + i\psi$ in Eq. then we find,

$$\phi + i\psi = r^n \cos n\vartheta + i r^n \sin n\vartheta,$$

whence,

$$\phi = r^{3/2} \cos 3/2\vartheta$$

$$\psi = r^{3/2} \sin 3/2\vartheta$$

for this example. The velocity components V_τ and V_ϑ are given by,

$$V_\tau = -\frac{\partial \phi}{\partial r}, \quad V_\vartheta = -\frac{1}{r}\frac{\partial \phi}{\partial \vartheta},$$

for Laplacian flow in general.

Hence at point 0 the velocity is zero, *i.e.* the point of intersection of the straight portions of the boundary represents a stagnation point in the flow. Eqs. indicate that the velocity increases with distance from the boundary, the magnitude simply being,

$$V\sqrt{V_\tau^2 + V_\vartheta^2} = \frac{3}{2} r^{1/2}.$$

Consequently, the solution can only apply over a finite distance from the boundary if the motion is to be physically real (*i.e.* have a finite velocity at infinity).

A thorough discussion of conformal transformations is beyond the scope of this text. For a more thorough treatment including a wide variety of solved problems of this type, reference is made to Milne-Thompson (1938).

3

Fluid Mechanics in Chemical Engineering

A knowledge of fluid mechanics is essential for the chemical engineer because them ajority of chemical -processing operation sarecon ducted eitherpartlyor totally in the fluid phase. Examples of such operations abound in the biochemical, chemical, energy, fermentation, materials, mining, petroleum, pharmaceuticals, polymer, and waste-processing industries.

There are two principal reasons for placing such an emphasis on fluids. First, at typical operating conditions, an enormous number of materials normally exist as gases or liquids, or can be transformed into such phases. Second, it is usually more efficient and cost-effective to work with fluids in contrast to solids. Even some operations with solids can be conducted in a quasi-fluidlike manner; examples are the fluidized-bed catalytic refining of hydrocarbons, and the long-distance pipelining of coal particles using water as the agitating and transporting medium.

Although there is inevitably a sign'ificant amount of theoretical development, almost all the material in this book has some application to chemical processing and other important practical situations. Throughout, we shall endeavor to present an understanding of the physical behaviour involved; only then is it really possible to comprehend the accompanying theory and equations.

GENERAL CONCEPTS OF A FLUID

We must begin by responding to the question, "What is a fluid?" Broadly speaking, a fluid is a substance that will deform continuously when it is subjected to a tangential or shear force, much as a similar type of force is exerted when a water-skier skims over the surface of a lake or butter is spread on a slice of bread. The rate at which the fluid deforms continuously depends not only on the magnitude of the applied force but also on a property of the fluid called its viscosity or resistance to deformation and flow. Solids will also deform when sheared, but a position of equilibrium is soon reached in which elastic forces induced by the deformation of the solid exactly counterbalance the applied shear force, and further deformation ceases.

A simple apparatus for shearing a fluid is shown in figure. The fluid is contained between two concentric cylinders; the outer cylinder is stationary, and the inner one (of radius R) is rotated steadily with an angular velocity É. This shearing motion of a fluid can continue indefinitely, provided that a source of energy—supplied by means of a torque here—is available for rotating the inner cylinder. The diagram also shows the resulting velocity profile; note that the velocity in the direction of rotation varies from the peripheral velocity RÉ of the inner cylinder down to zero at the outer stationary cylinder, these representing typical no-slip conditions at both locations. However, if the intervening space is filled with a solid—even one with obvious elasticity, such as rubber—only a limited rotation will be possible before a position of equilibrium is reached, unless, of course, the torque is so high that slip occurs between the rubber and the cylinder.

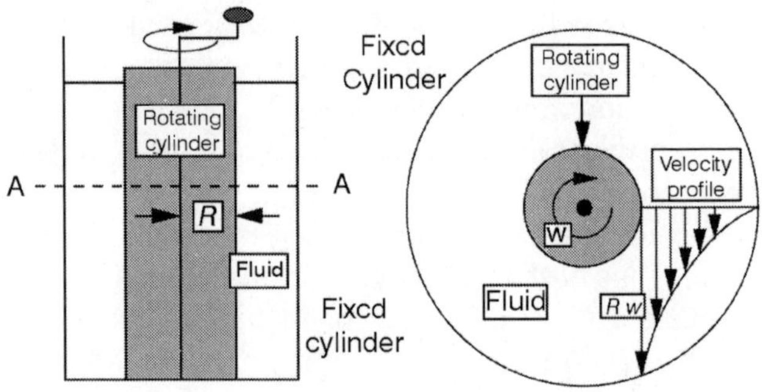

(a) Side elevation (b) Plan of section across A-A (not to scale)

Fig. Shearing of a fluid.

There are various classes of fluids. Those that behave according to nice and ob-vious simple laws, such as water, oil, and air, are generally called Newtonian fluids.

These fluids exhibit constant viscosity but, under typical processing conditions, virtually no elasticity. Fortunately, a very large number of fluids of interest to the chemical engineer exhibit Newtonian behaviour, which is devoted to the study of non-Newtonian fluids.

A fluid whose viscosity is not constant (but depends, for example, on the intensity to which it is being sheared), or which exhibits significant elasticity, is termed non-Newtonian. For example, several polymeric materials subject to defor-mation can "remember" their recent molecular configurations, and in attempting to recover their recent states, they will exhibit elasticity in addition to viscosity. Other fluids, such as drilling mud and toothpaste, behave essentially as solids and will not flow when subject to small shear forces, but will flow readily under the influence of high shear forces.

Fluids can also be broadly classified into two main categories—liquids and gases. Liquids are characterized by relatively high densities and viscosities, with molecules close together; their volumes tend to remain constant, roughly independent of pressure, temperature, or the size of the vessels containing them.

Gases, on the other hand, have relatively low densities and viscosities, with molecules far apart; generally, they will rapidly tend to fill the container in which they are placed. However, these two states—liquid and gaseous—represent but the two extreme ends of a continuous spectrum of possibilities.

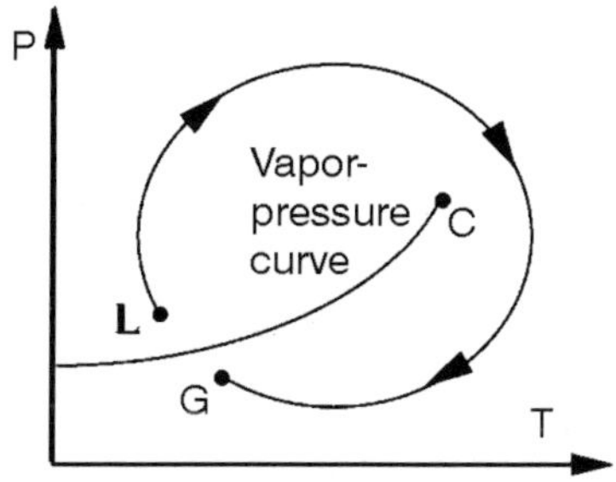

Fig. When does a liquid become a gas?

The situation is readily illustrated by considering a fluid that is initially a gas at point G on the pressure/temperature. By increasing the pressure, and perhaps lowering the temperature, the vapor-pressure curve is soon reached and crossed, and the fluid condenses and apparently becomes a liquid at point L. By continuously adjusting the pressure and temperature so that the clockwise path is followed, and circumnavigating the critical point C in the process, the fluid is returned to G, where it is presumably once more a gas. But where does the transition from liquid at L to gas at G occur? The answer is at no single point, but rather that the change is a continuous and gradual one, through a whole spectrum of intermediate states.

STRESSES, PRESSURE, VELOCITY, AND THE BASIC LAWS

Stresses. The concept of a force should be readily apparent. In fluid mechanics, a force per unit area, called a stress, is usually found to be a more convenient and versatile quantity than the force itself. Further, when considering a specific surface, there are two types of stresses that are particularly important.

- The first type of stress, acts perpendicularly to the surface and is therefore called a normal stress; it will be tensile or compressive, depending on whether it tends to stretch or to compress the fluid on which it acts. The normal stress equals F/A, where F is the normal force and A is the area of the surface on which it acts. The dotted outlines show the volume changes caused by deformation. In fluid mechanics, pressure is usually the most important type of compressive stress.

- The second type of stress, acts tangentially to the surface; it is called a shear stress τ, and equals F/A, where F is the tangential force and A is the area on which it acts. Shear stress is transmitted through a fluid by interaction of the molecules with one another. A knowledge of the shear stress is very important when studying the flow of viscous Newtonian fluids. For a given rate of deformation, measured by the time derivative $d\gamma/dt$ of the small angle of deformation γ, the shear stress τ is directly proportional to the viscosity of the fluid.

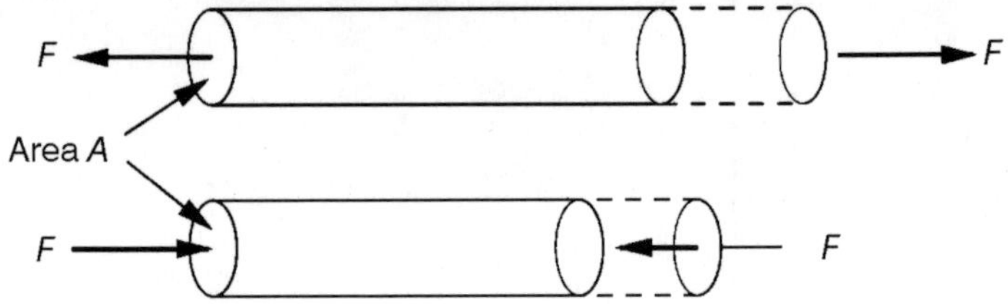

Fig. (a) Tensile and compressive normal stresses F/A, act-ing on a cylinder, causing elongation and shrinkage, respectively.

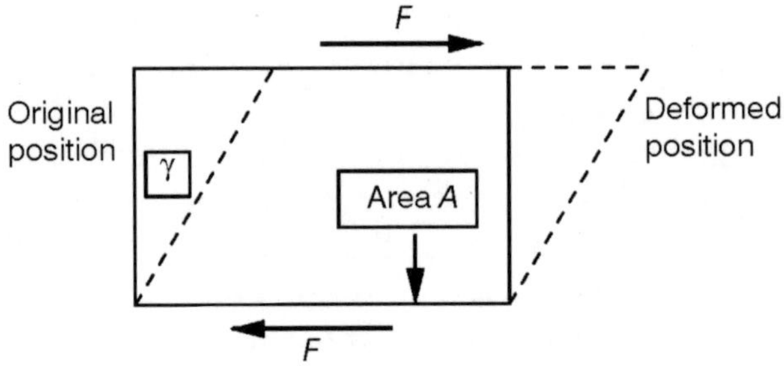

Fig. (b) Shear stress $\tau = F/A$, acting on a rectangular parallelepiped, shown in cross section, causing a deformation measured by the angle γ (whose magnitude is exaggerated here).

Pressure: In virtually all hydrostatic situations—those involving fluids at rest—the fluid molecules are in a state of compression.

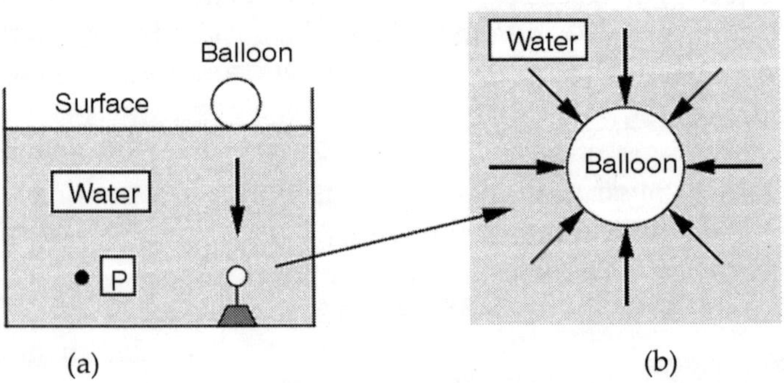

Fig. (a) Balloon submerged in a swimming pool; (b) enlarged view of the compressed balloon, with pressure forces acting on it.

For example, for the swimming pool whose cross section, this compression at a typical point P is caused by the downwards gravitational weight of the water above point P. The degree of compression is measured by a scalar, p—the pressure.

A small inflated spherical balloon pulled down from the surface and tethered at the bottom by a weight will still retain its spherical shape (apart from a small distortion at the point of the tether), but will be diminished in size. It is apparent that there must be forces acting normally inward on the surface of the balloon, and that these must essentially be uniform for the shape to remain spherical.

Although the pressure p is a scalar, it typically appears in tandem with an area A (assumed small enough so that the pressure is uniform over it). By definition of pressure, the surface experiences a normal compressive force $F = pA$. Thus, pressure has units of a force per unit area—the same as a stress.

The value of the pressure at a point is independent of the orientation of any area associated with it, as can be deduced with reference to a differentially small wedge-shaped element of the fluid.

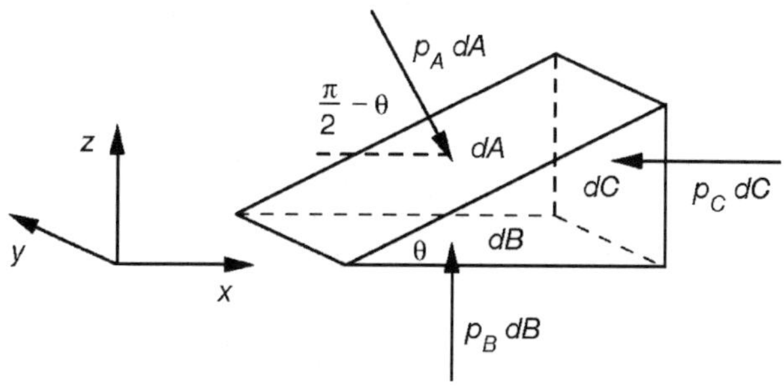

Fig. Equilibrium of a wedge of fluid.

Due to the pressure there are three forces, $p_A dA$, $p_B dB$, and $p_C dC$, that act on the three rectangular faces of areas dA, dB, and dC. Since the wedge is not moving, equate the two forces acting on it in the horizontal or x direction, noting that $p_A dA$ must be resolved through an angle $(\pi/2 - \theta)$ by multiplying it by $\cos(\pi/2 - \theta) = \sin\theta$:

$$p_A dA \sin\theta = p_C dC.$$

The vertical force $p_B dB$ acting on the bottom surface is omitted from Eqn. because it has no component in the x direction. The horizontal pressure forces acting in the y direction on the two triangular faces of the wedge are also omitted, since again these forces have no effect in the x direction. From geometrical considerations, areas dA and dC are related by:

$$dC = dA \sin\theta.$$

These last two equations yield:

$$p_A = p_C,$$

verifying that the pressure is independent of the orientation of the surface being considered. A force balance in the z direction leads to a similar result, $p_A = p_B$.

For moving fluids, the normal stresses include both a pressure and extra stresses caused by the motion of the fluid.

The amount by which a certain pressure exceeds that of the atmosphere is termed the gauge pressure, the reason being that many common pressure gauges are really differential instruments, reading the difference between a required pressure and that of the surrounding atmosphere. Absolute pressure equals the gauge pressure plus the atmospheric pressure.

Velocity. Many problems in fluid mechanics deal with the velocity of the fluid at a point, equal to the rate of change of the position of a fluid particle with time, thus having both a magnitude and a direction. In some situations, particularly those treated from the macroscopic viewpoint, it sometimes suffices to ignore variations of the velocity with position. In other cases—particularly those treated from the microscopic viewpoint, it is invariably essential to consider variations of velocity with position.

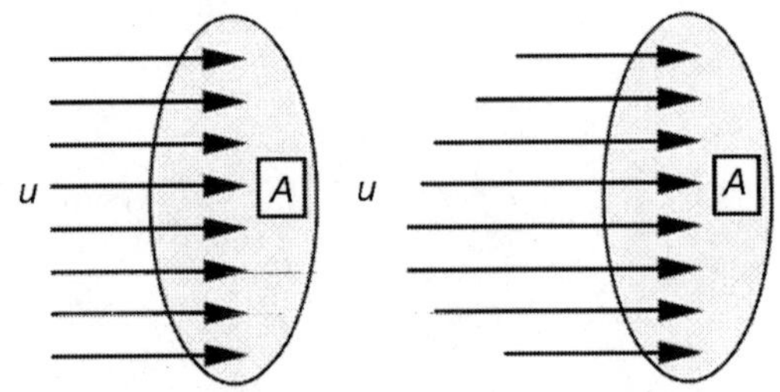

Fig. Fluid passing through an area A: (a) Uniform velocity, (b) varying velocity.

Velocity is not only important in its own right, but leads immediately to three fluxes or flow rates. Specifically, if u denotes a uniform velocity (not varying with position):

- If the fluid passes through a plane of area A normal to the direction of the velocity, the correspond-ing volumetric flow rate of fluid through the plane is $Q = uA$.
- The corresponding mass flow rate is $m = \rho Q = \rho uA$, where ρ is the (constant) fluid density. The alternative notation with an overdot, $\dot{m}$, is also used.
- When velocity is multiplied by mass it gives momentum, a quantity of prime importance in fluid mechanics. The corresponding momentum flow rate pass-ing through the area

A is $\dot{M} = mu = \rho u^2 A$.

pressions will be seen later to involve integrals over the area A: $Q =$

$\int_A u\,dA, m \int_A \rho u^2 dA.$

Basic laws. In principle, the laws of fluid mechanics can be stated simply, and—in the absence of relativistic effects—amount to conservation of mass, energy, and momentum. When applying these laws, the procedure is first to identify a system, its boundary, and its surroundings; and second, to identify how the system interacts with its surroundings. Let the quantity X represent either mass, energy, or momentum. Also recognize that X may be added from the surroundings and transported into the system by an amount X_{in} across the boundary, and may likewise be removed or transported out of the system to the surroundings by an amount X_{out}.

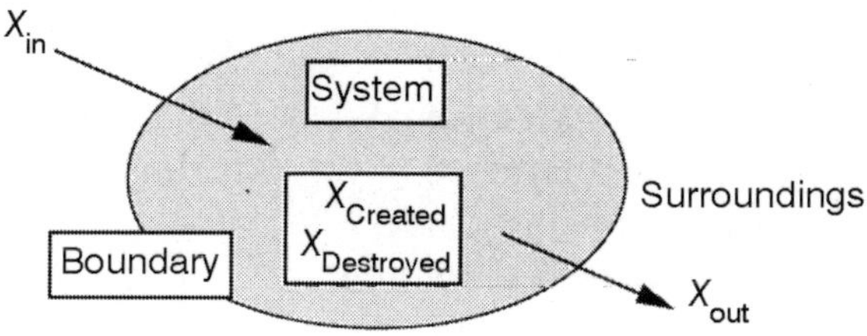

Fig. A system and transports to and from it.

The general conservation law gives the increase "X_{system} in the X-content of the system as:

$$Xin - X_{out} = \Delta X_{system}.$$

Although this basic law may appear intuitively obvious, it applies only to a very restricted selection of properties X. For example, it is not generally true if X is another extensive property such as volume, and is quite meaningless if X is an intensive property such as pressure or temperature.

In certain cases, where X^i is the mass of a definite chemical species i, we may also have an amount of creation $X^i_{created}$ or destruction $X^i_{destroyed}$ due to chemical reaction, in which case the general law becomes:

$$X^i_{in} - X^i_{out} + X^i_{created} - X^i_{destroyed} = \Delta X^i_{system}.$$

The conservation law, and such fundamental importance that in various guises it will find numerous applications throughout all of this text.

To solve a physical problem, the following information concerning the fluid is also usually needed:

- The physical properties of the fluid involved.
- For situations involving fluid flow, a constitutive equation for the fluid, which relates the various stresses to the flow pattern.

FLUID MECHANICS

Continuum Mechanics is a means of studying the deformation and flow of matter by ignoring its particulate nature. It is a simplification that makes it possible to investigate the movement of matter on scales larger compared to typical distances between particles. Thus, in studying the movement of air in the atmosphere the fact that is it made up of atoms is ignored. It would, in any case, be impossible to predict the motion of each atom. Fortunately, when asking questions such as "How warm will it be tomorrow?" we are only interested in the average behaviour of large numbers of atoms, not in their individual motion. Similarly, in the study of mud flow the movement of industrial clay particles is ignored. At larger scales the evolution of galaxies can also be described using continuum models instead of trying to predict the movement of every planet, asteroid, and star.

This approach is used to study a huge range of phenomena, including the flow of air and water as well as in studies of rock slides (the individual particles being large rocks), snow avalanches, blood flow, and even galaxy evolution.

The Continuum and Fluid Mechanics courses in the Applied Mathematics department at the University of Waterloo. We hope that the material contained in these pages enriches the experience of those taking our courses (AM 361 and AM 463). At the same time, for those who are at an earlier point in their university education, we hope to show you both the richness of our field of study as well as illustrate how the courses you take earlier in your career tie in with what we do in our upper year courses and in our research.

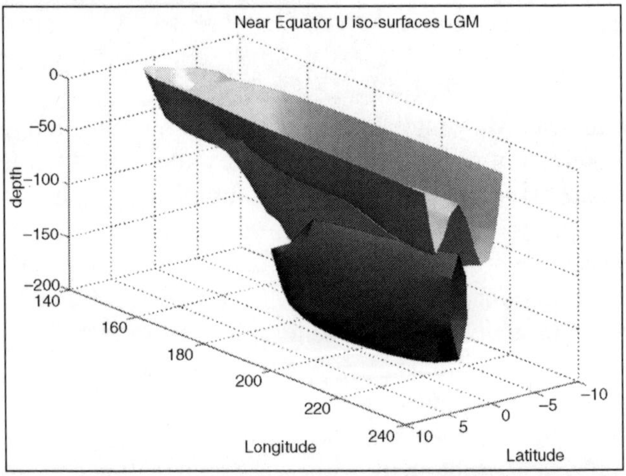

Continuum and Fluid Mechanics each have a long history. Indeed we take some past achievements nearly for granted whenever we fly in an airplane or turn on the television to check the weather report. Yet it is likely true that the wide availability of computers for research purposes makes the present a new Golden Age for the study of both fluid and continuum mechanics. It is perhaps easier to define fluid mechanics than continuum mechanics. After all we are surrounded, indeed bathed in, fluids for the entirety of our lives. This is because at the level of their gross, or macroscopic, behaviour (how they flow, for example) both gases and liquids can be considered to be fluids.

However, the actual mathematical description of fluid flow can be vastly different depending on the scales of interest. Thus, while the aerodynamics of a bumble bee's flight and the evolution of a hurricane are technically governed by the same set of equations, the simplifications made to reflect the dominant physics of each situation are wildly different.

As broad as fluid mechanics is, continuum mechanics can be considered to be even broader. This is because it is concerned with the general theory of equations that govern fluids, solids and all the exotic materials in between, from a general, and often abstract, point of view. Moreover, since much of the materials that surround us and make much of modern life possible (plastics being the most obvious example) are neither fluid nor solid, but something in between, there is also an intensely practical side to continuum mechanics. Indeed, the study of materials ranges across many disciplines in both the pure and applied sciences. One of the things we hope to convey throughout these web materials is the particular nature of studying materials within an applied mathematics setting.

VORTICES

Vortices are common features in fluid flows, They have important implications for the flight of airplanes where they are generated by boundary layer processes. Wing tip vortices are a limiting factor for the frequency with which airplanes can take off. An example of a much larger scale vortex is a hurricane. Flow at these scales is strongly affected by the earth's rotation: the flow around a hurricane is counterclockwise in the northern hemisphere and clockwise in the southern hemisphere.

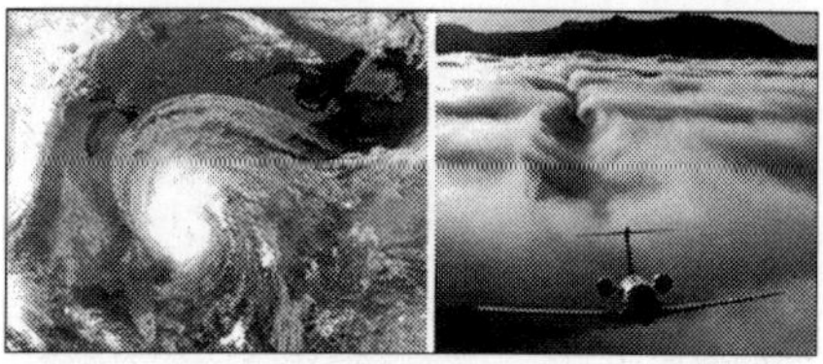

Fig. CessnaDownwash. Photo Credit: Gallery of Fluid Motion, Physics of Fluids A, Vol. 5, September 1993. Contribution by Hiroshi Higuchi (Syracuse University).

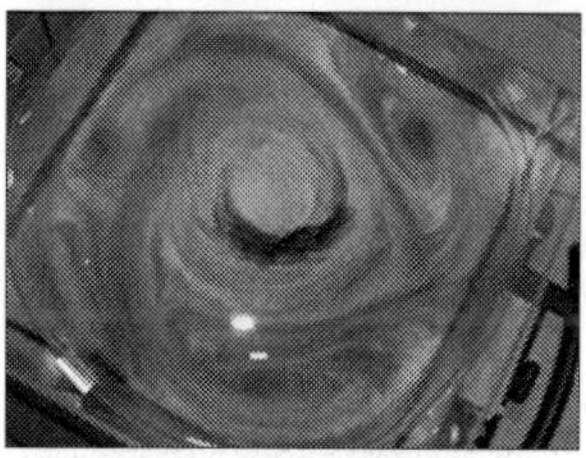

INTERNAL GRAVITY WAVES

Internal gravity waves are waves that occur in density stratified fluids in the presence of a gravitational field. They arise as gravitational restoring forces act on vertically displaced fluid. Internal gravity waves are all around us, but difficult to see. They are ubiquitous in stratified lakes, in the ocean, and in the atmosphere. They are easily seen in a glass jar containing cooking oil and water. Internal waves are the waves you see on the interface. Better yet, try a layer of fresh water over salty water dyed with food colouring.

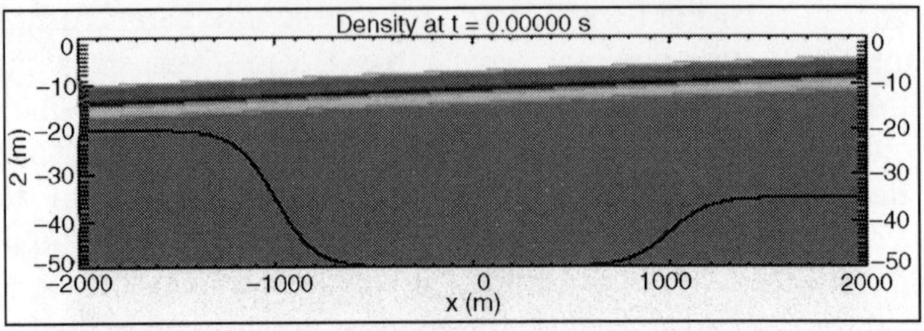

This animation shows how internal waves can be generated in a lake. Daytime winds drag warm surface water down wind forming a tilted epilimnion, or thermocline (the thin layer across which the water temperature changes). When the winds die down, the thermocline is released. The rocking basin scale standing wave, or seiche, evolves into a complex wave field, including trains of non-linear solitary waves.

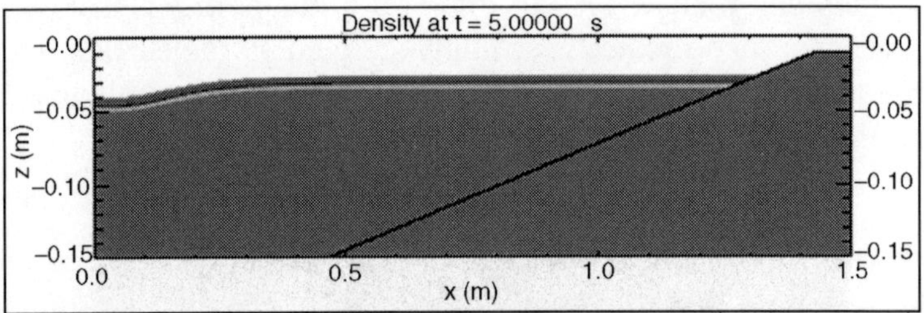

When a single solitary wave shoals along a sloping bottom it breaks, dissipating energy, mixing fluid and resuspending bottom sediments. This has important implications for biological and chemical processes.

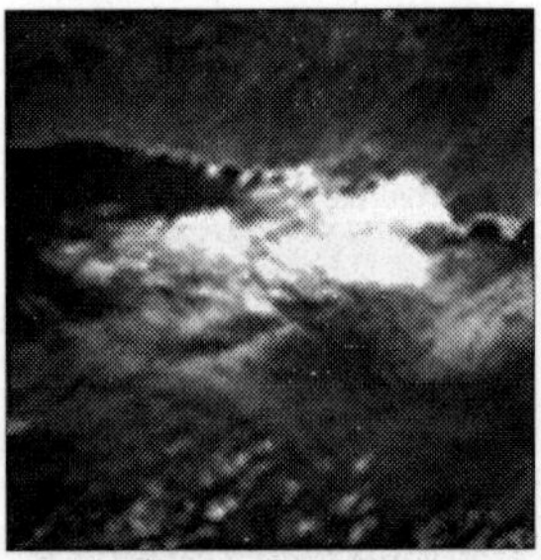

Internal Waves in the ocean and lakes lie beneath the surface. Their associated currents can modify the surface wave field, creating slicks on the surface which are visible from land or air. Here is a photo of internal wave slicks taken from the space shuttle.

Large amplitude internal waves in the ocean are usually generated by tidal flow over topographic features such as a bank edge.

CONCEPT OF FLUID MECHANICS

Fluid Mechanics is a sub discipline of Continuum Mechanics which focuses on the behaviour of fluids which include liquids and gases. Important examples for our everyday lives are water and air.

Beautiful photos of fluids in motion can be found in the Physics of Fluid Gallery. Some of our own images and discussions of fluids dynamics from a student perspective can be found here. Fluid mechanics is a branch of physics which concerns the study of fluids and the ways in which they interact with forces. Both liquids and gases are considered to be fluids for the purposes of this branch of science. Often, the field of fluid mechanics is divided into two more specific fields of study. These are fluid statics and fluid dynamics, which concern fluids at rest and fluids in motion, respectively. Fluid mechanics can involve highly complex mathematics, and the aid of modern computers has enhanced this science significantly.

The chronological roots of fluid mechanics go all the way back to at least the ancient Greeks. The Greek physicist and inventor Archimedes was the author of some of the first studies we know of which concern fluid statics, including the property of buoyancy. Persian philosophers in the Medieval time period coupled these ancient works with their own studies of fluid dynamics that acted as an early precursor to modern fluid dynamics. Such well-known historical figures as Leonardo da Vinci and Sir Isaac Newton, as well as others, made notable contributions to our understanding of fluid mechanics.

Every type of science starts out with basic, fundamental assumptions that govern the course of their study. Fluid mechanics is typically defined as having three basic premises or assumptions at its root. The first is the conservation of mass, which means that mass can neither be spontaneously created nor destroyed, although it may change forms. The second assumption, the conservation of momentum, is somewhat similar. This law states that the total momentum in a closed system is constant, and cannot spontaneously appear or disappear.

The third basic assumption governing fluid mechanics is what is known as the continuum hypothesis. This is a way of seeing fluids that does not take into account the presence of discrete molecules. Instead, a fluid's properties are assumed to vary in a continuous way from one point to the next.

Because it ignores the actual nature of small particles of matter, the continuum hypothesis is only an approximation used as a tool in calculations. It can result in a slightly inaccurate solution, but also in solutions that are very accurate under ideal circumstances. Other, more exact methods exist, but this hypothesis is often quite useful as a preliminary assumption. Many times, it can also be assumed that a given fluid is incompressible, meaning it cannot be compressed. This is only actually true of liquids, however, and not gases.

- Our courses Continuum Mechanics (AMATH 361) and Fluid Mechanics (AMATH 463) provide a unique opportunity for undergraduates to learn about this fascinating field. These courses combine mathematical analysis of the physical phenomena with computer visualization and demonstrations using equipment in our Fluids Lab
- The option in Earth Science, which can be combined with our AMATH programme, is designed for students who are interested in Fluid Mechanics and would like to learn how it is related to environmental issues.

Control Theory

- We have a fourth year course in Control Theory (AMATH 455), which builds on the foundational courses in ordinary differential equations and complex analysis. We also have a related course in the Calculus of Variations (AMATH 456).
- When taking the control theory course you will do some experiments in our Control Lab.
- If you find Control Theory interesting, consider enrolling in our programme Applied Mathematics with Engineering Electives. One of the areas of specialization is "Communication and Control". You could take a course in "Robot Dynamics and Control".

DYNAMICAL SYSTEMS

- Two of our foundation courses (AMATH 250 and AMATH 351) deal with methods for analyzing the solutions of differential equations. They provide the background for a fourth year course on dynamical systems.

Quantum Theory

- An introduction to the fascinating quantum world is provided by our courses AMATH 373/473 (Quantum Theory 1 and 2).
- AMATH 473, together with PHYS 467 (Introduction to Quantum Information Processing), will provide the foundation for those interested in research in Quantum Information Theory and Quantum Computing.
- If you find quantum theory interesting, consider enrolling in our programme in Mathematical Physics.

Scientific Computing

- Our courses AMATH 342 (Computational methods for DEs) And AMATH 442 (Computational methods for PDEs) give a solid introduction to the development of computational methods, their implementation in Matlab, and numerical simulations in a variety of areas of applications.

- If using computers to solve problems in science and technology appeals to you consider enrolling in our programme Scientific Computation/Applied Mathematics.

Have you ever wondered how airplanes fly (mathe-maticians 'proved' they could not around the turn of the 19th century!) or how weather predictions are made? Why dimples on a golf ball increase the distance it travels? Why curve balls curve? Do toilets really flush counterclockwise in the northern hemisphere and clockwise in the southern hemisphere? Why are surface water waves almost always nearly parallel to the beach? These are questions you can learn answers to if you study Applied Mathematics at Waterloo.

ENERGY IN MECHANICS

MECHANICAL ENERGY

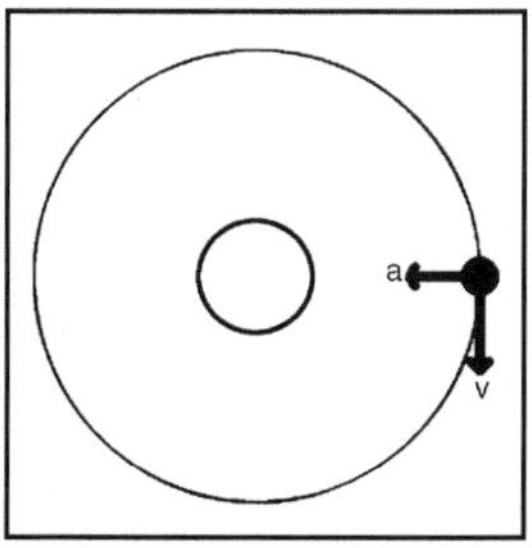

Figure: A satellite is orbiting the Earth only influenced by the conservative gravitational force and the mechanical energy is therefore conserved. The satellite is accelerated towards the Earth with an acceleration perpendicular to the orbit. This acceleration is represented by a green acceleration vector and the velocity is represented by a red velocity vector. Though the velocity is constantly changed with the direction of the vector because of the acceleration vector, the speed of the satellite is not since the magnitude of the velocity vector remains unchanged.

In physics, mechanical energy is the sum of potential energy and kinetic energy present in the components of a mechanical system. It is the energy associated with the motion and position of an object. The law of conservation of energy states that in an isolated system that is only subject to conservative forces, like the gravitational force, the mechanical energy is constant. If an object is moved in the opposite direction of a conservative net force, the potential energy will increase and if the speed (not the velocity) of the object is changed, the kinetic energy of the object is changed as well. In all real systems, however, non-conservative forces, like frictional forces, will be present, but often they are of negligible values and the mechanical energy's being constant can therefore be a useful approximation. In elastic collisions, the mechanical energy is conserved but in inelastic collisions, some mechanical energy is converted into heat. The equivalence between lost mechanical energy and an increase in temperature was discovered by James Prescott Joule.

Many modern devices, such as the electric motor or the steam engine, are used today to convert mechanical energy into other forms of energy, *e.g.* electrical energy, or to convert other forms of energy, like heat, into mechanical energy.

General

Energy is a scalar quantity and the mechanical energy of a system is the sum of the potential energy which is measured by the position of the parts of the system, and the kinetic energy which is also called the energy of motion:

$$E_{mechanical} = U + K$$

The law of conservation of mechanical energy states that if a body or system is subjected only to conservative forces, the total mechanical energy of that body or system remains constant. The difference between a conservative and a non-conservative force is that when a conservative force moves an object from one point to another, the work done by the conservative force is independent of the path. On the contrary, when a non-conservative force acts upon an object, the work done by the non-conservative force is dependent of the path.

The potential energy, *U*, depends on the position of an object subjected to a conservative force. It is defined as the object's ability to do work and is increased as the object is moved in the opposite direction of the direction of the force. If *F* represents the conservative force and × the position, the potential energy of the force between the two positions x_1 and x_2 is defined as the negative integral of *F* from x_1 to x_2:

$$U = -\int_{x_1}^{x_2} \vec{F} \cdot d\vec{x}$$

The kinetic energy, *K*, depends on the speed of an object and is the ability of a moving object to do work on other objects when it collides with them. It is defined as one half the product of the object's mass with the square of its speed, and the total kinetic energy of a system of objects is the sum of the kinetic energies of the respective objects:

$$K = \frac{1}{2} m v^2.$$

Conservation and Interconversion of Energy

Of the three great conservation laws of classical mechanics, the conservation of energy is regarded as the most important. According to this law, the mechanical energy of an isolated system remains constant in time, as long as the system is free of all frictional forces, including eventual internal friction from collisions of the objects of the system.

In any real situation, frictional forces and other non-conservative forces are always present, but in many cases their effects on the system are so small that the principle of conservation of mechanical energy can be used as a fair approximation.

Though energy cannot be created nor destroyed in an isolated system, it can be internally converted to any other form of energy.

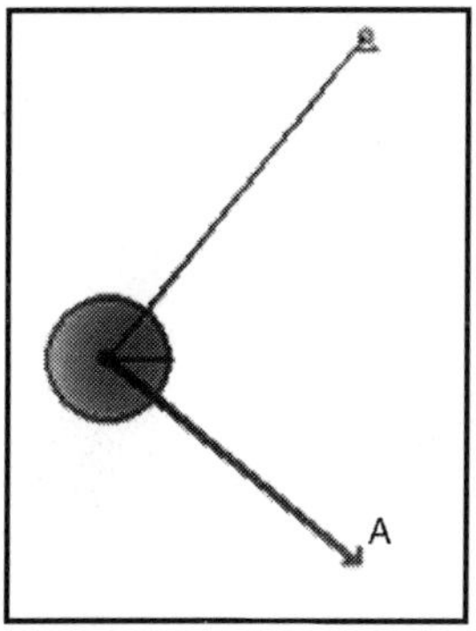

Figure: A swinging pendulum with the velocity vector (green) and acceleration vector (blue). The magnitude of the velocity vector, the speed, of the pendulum is greatest in the vertical position and the pendulum is farthest from Earth in its extreme positions.

Thus, in a mechanical system like a swinging pendulum subjected to the conservative gravitational force where frictional forces like air drag and friction at the pivot are negligible, energy passes back and forth between kinetic and potential energy but never leaves the system.

The pendulum reaches greatest kinetic energy and least potential energy when in the vertical position, because it will have the greatest speed and be nearest the Earth at this point. On the other hand, it will have its least kinetic energy and greatest potential energy at the extreme positions of its swing, because it has zero speed and is farthest from Earth at these points. However, when taking the frictional forces into account, the system loses mechanical energy with each swing because of the work done by the pendulum to oppose these non-conservative forces.

That the loss of mechanical energy in a system always resulted in an increase of the system's temperature has been known for a long time, but it was the amateur physicist James Prescott Joule who first experimentally demonstrated how a certain amount of work done against friction resulted in a definite quantity of heat which should be conceived as the random motions of the particles that matter is composed of. This equivalence between mechanical energy and heat is especially important when considering colliding objects. In an elastic collision, mechanical energy is conserved; *i.e.* the sum of the kinetic energies of the colliding objects is the same before and after the collision.

After an inelastic collision, however, the total mechanical energy of the system will have changed. Usually, the total mechanical energy after the collision is smaller than the initial total mechanical energy and the lost mechanical energy is converted into heat.

However, the total mechanical energy can be greater after an inelastic collision if for example the collision causes an explosion which converts chemical energy into mechanical energy.

In inelastic collisions, the smaller particles of which the colliding objects consist are shaken up and rattle around. These small-scale motions are perceived as an increase in heat and need kinetic energy which must be taken from the large-scale motion of the objects that are observed directly. Thus, the total energy of the system remains unchanged though the mechanical energy has been changed.

ELEMENTARY MECHANICS AND MACHINES

We now know enough to consider the operation of simple machines, such as levers, ramps, gears, pulleys, screw jacks, and so on. An understanding of these devices is based on elementary concepts of force, work, and power. A lever is one of the simplest tools. Suppose Dexter wants to lift a weight or "load" using a long stout stick as a lever.

He props the stick on top of a rock for support so that three-quarters of the stick is on his side of the rock, while the other quarter is on the other side of the rock and wedged under the load. The rock is said to act as a "fulcrum", by the way. If he pulls down on the lever, his end of the stick moves three times as far down as the other end of the stick under the load moves up. The total work done on both ends of the stick is the same, but since Dexter's end of the stick moves three times farther, he only ends up applying a third of the force. This means Dexter has a "mechanical advantage (MA)" of three to one, or in more formal terms:

$$\text{Force}_1 \times \text{displacement}_1 = \text{force}_2 \times \text{displacement}_2$$

To balance a lever, the product of the mass and length of the lever on one side of the fulcrum must equal the product of the mass and length of the lever on the other side of the fulcrum:

$$\text{mass}_1 \times \text{length}_1 = \text{mass}_2 \times \text{length}_2$$

In this example, if the lever is a meter long and the fulcrum is 25 centimeters from one end, then 1 kilogram at the long end of the fulcrum balances 3 kilograms at the short end. If the fulcrum is 20 centimeters from one end, then 1 kilogram will balance 4 kilograms. Similar issues of balance comes up in other domains of physics.

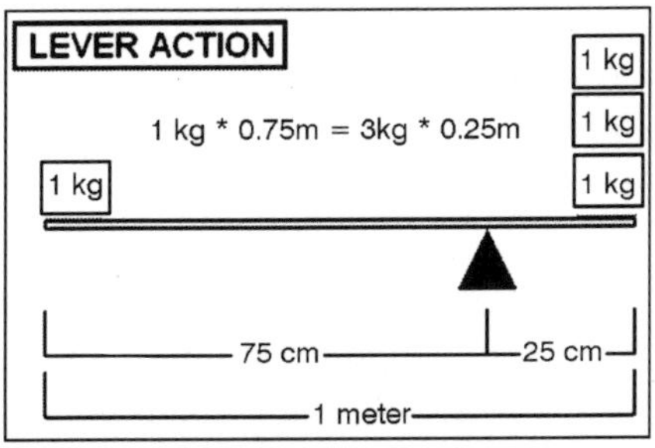

The example above involves arranging a lever to use a small force over a long distance to exert a big force over a short distance. Such a lever configuration can be reversed, using a big force over a short distance to exert a small force over a long distance; this might seem pointless at first sight, but it is useful in building a catapult.

The Medieval catapult known as a "trebuchet" (pronounced "tray-byoo-shay" though sometimes Anglicized to "tray-bucket") consists of a heavy weight connected to the short lever arm and a sling for a stone projectile connected to the long lever arm. When released, the heavy weight moves the short lever arm over an arc, with the long lever arm moving over the same arc but covering a longer distance, meaning the sling moves the payload move at a higher velocity.

The increase in velocity is the inverse of the mechanical advantage: if the long lever arm is, say, five times longer than the short lever arm, the sling and its payload has five times the velocity of the short lever arm. This configuration of lever, incidentally, is known as a "first class lever", meaning that the load is on one end, the work is exerted on the other end, and the fulcrum is between the two. There are two other lever configurations. The "second class lever" features the fulcrum on one end, the work on the other end, and the load between the two: it's a good scheme for nutcrackers, manual paper punches, and the little wall-mounted press I use to crush aluminum cans for recycling.

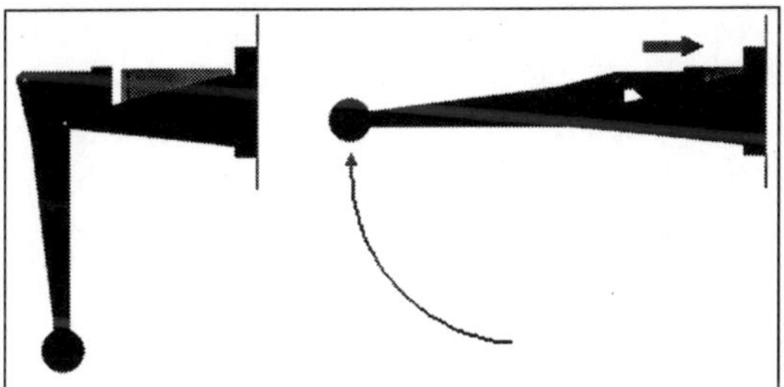
Second Order Lever: A Can Press

The concept of mechanical advantage applies to the second class lever basically as it does to a first class lever. If Dexter has a nutcracker with the jaws clamping down on a walnut a third of the way from the hinge to the grip point, he still gets a mechanical advantage of three to one.

However, a "third class lever" which features the fulcrum on one end, the load on the other end, and the work exerted in between always has a mechanical advantage of less than 1, meaning the force applied to the load is always less than force exerted on the lever. In reality, a third class lever is not used to obtain a mechanical advantage as such.

It is used to extend reach, for example in ice or sugar cube or salad tongs. It can also be used as a type of catapult mechanism: if a force is applied near the base of a pivoting arm, the tip of the arm will move over a longer path in the same amount of time, allowing it to throw off a load at high velocity.

Third-Order Lever (GVG/PD)

A ramp is another simple machine. Suppose Dexter wants to lift a load one meter into the back of a truck. Lifting it straight up is a good way to hurt himself, so he finds a stout plank three meters long and rolls the load up the plank on a dolly. Once more he only does a third of the work for three times the distance, and has a mechanical advantage of three to one.

A screw jack is simply a ramp wrapped around an axle, with the axle rotated by a handle used as a lever. If the circular motion of the handle is 30 centimeters per revolution, for example, and one turn of the screw lifts a weight by one centimeter, then the mechanical advantage is 30 to 1.

My little Toyota car has a scissors-type screw jack, in which rotating the handle turns a horizontal screw that raises a mechanical scissors arrangement. Obviously a detailed analysis of this device is a little more complicated, but a simple "input-output" analysis shows that if one rotation of the handle travels 30 centimeters to raise the jack one centimeter, I have a mechanical advantage of 30 to 1.

There are other ways to obtain "leverage". For example, a windlass consists of a crank with a radius Rc turning an axle with a radius Ra to wind up a rope. One turn of the crank means one turn of the axle, and so the mechanical advantage is Rc / Ra.

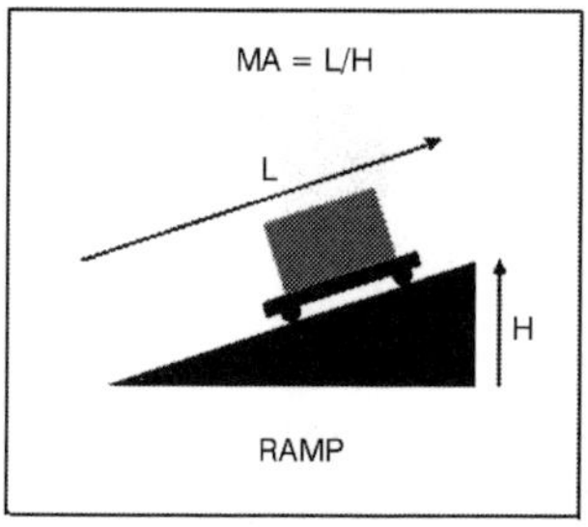

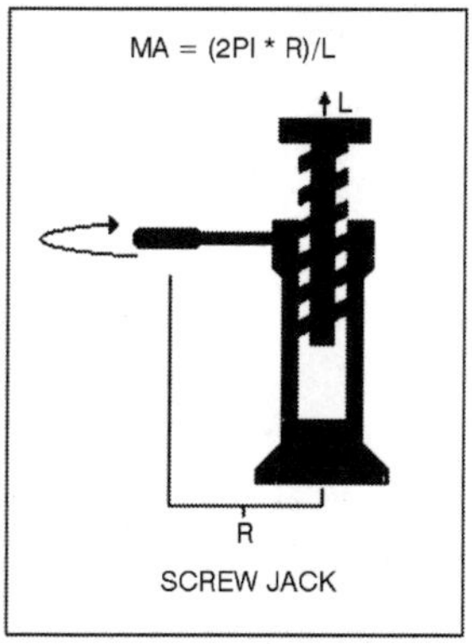

A simple pulley doesn't provide any mechanical advantage, it merely allows a force to be applied in a more convenient direction. For example, a rope slung over a pulley might be tied to the hitch of a car, with the car then driving forward to lift a weight off the ground. A block and tackle involves a pair of pulleys looped together.

Pulling on the rope out of one end of the loop makes the loop shorter, but as the loop has two sides, it gets shorter at only half the rate the rope is pulled out of the loop. This means a mechanical advantage of two to one. If Dexter makes a block and tackle with two loops, he gets a mechanical advantage of four to one; if he makes it with three loops, and he gets an MA of six to one.

Gears are probably the best-known of all simple machines. Two gears meshing together are basically just a set of wheels moving against each other, with the teeth ensuring they don't slip. If the driving gear is only a fifth as big in diameter as the gear it is driving, then the driving gear performs five rotations for every rotation of the driven gear. or has a "gear ratio" of 5:1. This also means an MA of 5:1.

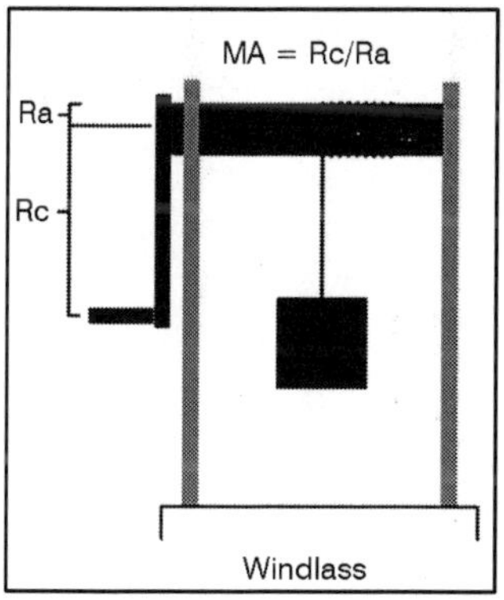

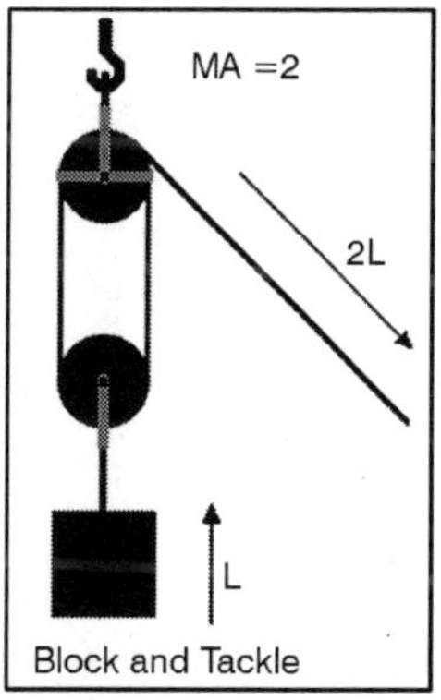

A classic manual or "stick" automotive transmission is no more than a box full of gears of different ratios. For cruising on level ground, a high gear ratio is used, providing high rolling speed with a minimum of power. For climbing up a hill, a low gear ratio is used, providing low rolling speed with a maximum of power. Of course the transmission has a gear scheme that allows putting the wheels in reverse.

The transmission is linked to the engine through a spring-loaded friction plate or "clutch". If Dexter is wants to change the gear ratio, he steps on a clutch pedal to release the clutch plate from the engine driveshaft, which temporarily "unloads" the transmission. He then changes the gear ratio with the stick and releases the clutch pedal, loading the transmission again.

Traditionally, manual transmissions have four to six gear ratios. Now many cars have a "continuously variable transmission (CVT)" that can vary continuously from its lowest gear ratio to its highest gear ratio. Actually, calling it a "gear ratio" is somewhat misleading, since a CVT doesn't have gears as such.

To get the basic idea of a CVT, imagine a small car with a transverse (sideways) mounted engine and front-wheel drive. There's a disk on the engine shaft and a disk on the front-wheel driveshaft, with the two disks linked only by a belt.

Obviously, if there's a disk with a small diameter on the engine shaft and a disk with a big diameter on the driveshaft, the gear ratio is low, good for climbing hills; if the engine shaft disk is big and the driveshaft disk small, then the gear ratio is high, good for economical fast cruise.

The magic trick in a CVT is to make the two disks effectively change diameter. This is done by actually making each in the form of *two* disks, in the shape of shallow cones with the tips together, one of the cones being fixed to the shaft and the other able to be moved up and down the shaft using an actuator system.

Pulling the two cones apart makes the belt ride lower in the space between the cones, reducing the diameter over which the belt is driven, while drawing the cones together makes the belt ride higher, increasing the diameter.

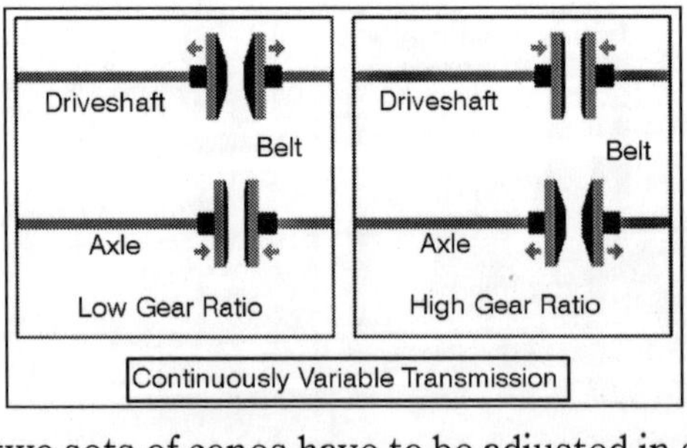

Continuously Variable Transmission

Of course, the two sets of cones have to be adjusted in step, one set closing and the other opening, to ensure that the belt doesn't go slack. The idea that a belt can be manufactured that will put up with this sort of workout is hard to believe; it consists of dual sets of steel bands, held in place with a string of strong metal cross members. The only shift settings necessary for a CVT are reverse, park, and forward. A CVT works best at relatively low speeds and in stop-and-go traffic. Due to friction losses in the belt, it offers no particular benefits at high speeds. The basic concept of the CVT is not new, going back to the Renaissance polymath Leonardo da Vinci at least. The first practical CVT design was patented by Dutch inventors Hub and Wim van Dorn in 1958, and the first full production cars with CVTs appeared in the late 1980s.

FORCES AND FIELDS

- We have defined force, and discussed gravitational forces a little. Gravity is actually only one of four fundamental forces in the Universe:
- Gravitational
- Electromagnetic
- Strong nuclear
- Weak nuclear

All other forces are just indirect manifestations of these four basic forces. The only ones we actually deal with directly are the gravitational and electromagnetic forces. The strong nuclear force holds the cores of atoms together, while the weak force only shows up in certain processes of atomic disintegration, and in fact it seems so unlike a force that physicists tend to call it the "weak interaction" instead.

The strong and weak nuclear forces were completely unknown to classical physicists and will not be mentioned further in this document.

The gravitational force is the most familiar of these four forces, since we deal with it all the time. Gravity is the weakest of all the four forces, and is about forty orders of magnitude (one followed by 40 zeroes) times weaker than the electromagnetic force. To visualize just how weak a force gravity is, consider that when Dexter picks up a brick, he is overcoming the gravitational attraction of the *entire Earth* on that brick. However, gravity is the only one that is effective over long distances, and so predominates over the entire Universe. Gravity is an attractive force between two or more masses, such as the Earth and the Moon. It's not actually restricted to such large masses, since the attraction exists between any two masses of any size, but since gravity is so weak it's hard to notice in smaller masses. As Newton discovered, the size of this force is proportional to the size of the two masses, and inversely proportional to the square of the distance between them:

$$\text{gravitational force (G)} = \text{constant} \times \text{mass}_1 \times \text{mass}_2 / \text{distance}^2$$

Where the constant has a value of 6.672×10^{-11} in standard metric units, and the distance is measured between centre of the individual masses. Double the size of one of the masses, and the force is multiplied by two. A big planet like Jupiter has much higher gravity than Earth, 2.54 times greater to be specific. Double the distance, the force is divided by four. If Dexter blasts off from Earth in a rocket, the farther he goes, the more the force falls off.

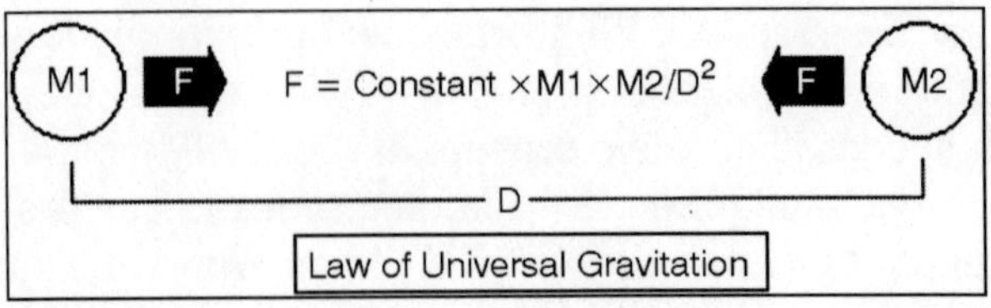

Dexter flies around the solar system to other planets all the time, experiencing different gravitational forces and having a different weight on each planet.

Since the vast majority of us never get very far off the ground, much less travel to other planets, in our practical experience the force of gravity is a constant, with an acceleration of 9.81 meters per second, as discussed earlier.

It should be noted that Newton never really claimed he understood *why* gravity behaved as it did, famously stating: "I frame no hypotheses." He accepted it as simply as a given, a fundamental rule of the way the Universe worked, and though modern physics has taken the idea somewhat farther, this document accepts it as a fundamental rule as well.

Physicists use the concept of a "field" to describe forces like gravity. This can be considered as a three-dimensional map of the direction and magnitude of the force at any point in space. The idea was originally conceived by the brilliant English researcher Michael Faraday.

Two-dimensional representations of fields usually use lines to map such fields, with the density of the lines giving the strength of the field and arrows attached to the lines to give their directions. Leveraging off this abstraction, physicists often refer to "lines of force" or "lines of flux" in descriptions of fields and field interactions, though in classical physics the fields are actually continuous and the lines are just a convenient abstraction, much like the altitude lines around mountains and other geographical features on relief maps.

The density of the lines of force gives the strength of the field and the amount of force it produces. In the special case of living on the surface of the Earth, the gravitational field is effectively uniform.

No matter where we place a mass in the three dimensions of our immediate surroundings, the direction of the force is straight down and the magnitude is effectively the same.

Potential energy in this uniform field, as mentioned in the previous section, is directly proportional to the height of a mass. If the mass is raised any significant fraction of the Earth's diameter, the gravitational force falls off and calculating potential energy is trickier.

Every object in the Universe exerts a force on every other object in the Universe. However, in most analyses of gravitational forces in the greater Universe, the gravitational field is regarded as acting between simple points of matter that have no extent. This is actually a very good approximation for nearly all practical purposes.

Visualizing this from a field perspective also helps explain why gravity obeys an "inverse square" law, decreasing in strength with the square of the distance. The gravitational force of the Earth can be visualized as lines of force radiating with uniform density in all directions, with the direction of the lines pointing into the Earth.

Now let's draw an invisible shell around the Earth at some radius. All the lines of force pass through this shell. Since the surface area of a sphere is 4 × πI × radius^2, then if we draw a second shell at double the radius, the surface area of this shell is four times that of the first.

The same number of lines of flux pass through this outer shell as they do the inner shell, but the surface area has been squared, meaning the density and so the strength of the gravitational field has been attenuated by the same factor.

Electromagnetic forces have similarities to and differences from gravitational forces. Gravitational forces occur between masses. Mass is irrelevant to electrical forces, they only rely on a property known as the "electrical charge" of the objects involved.

"Charge" is another fundamental concept, one that can only be explained in terms of itself: an electric field works on charges, charges are what an electrical field works on. However, charges are tangible, and can be measured. Physics defines a certain fundamental unit of charge that cannot be subdivided further.

This charge is very small and is usually measured in terms of 6.24E18 of these basic charges; this amount is known as a "coulomb". Charges can flow from one object to another, with a rate of flow of one coulomb per second given as an "ampere".

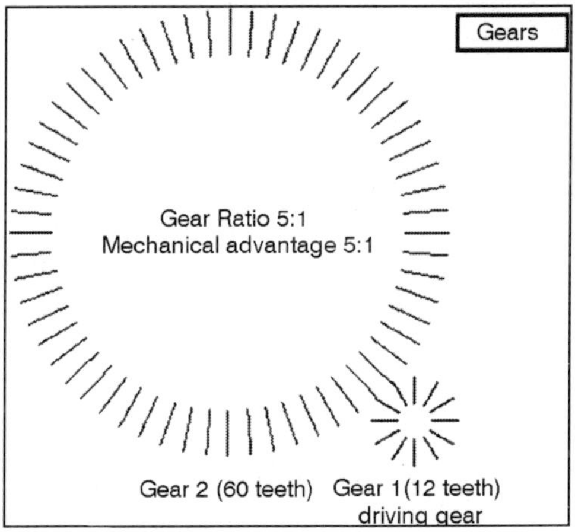

The basic law of electrical force is very similar to that of gravity. The electrical force is proportional to the product of the charges of each object and inversely proportional to the square of the distance between them:

Electrical force = constant × $charge_1$ × $charge_2$/$distance^2$

where the constant has a value of 8.9875E9 in standard metric units. A charged object in an electric field will move in the direction of the lines of force. This relationship was discovered by the pioneering French electrical researcher Charles Augustin de Coulomb, for whom of course the coulomb of charge is named, and the electrical force is sometimes called the "Coulomb force" in his honour.

Dexter can easily demonstrate this force by rubbing a balloon on his hair, a process that strips off charges; the balloon will then stick to walls or the ceiling. The same effect makes styrofoam packing peanuts difficult to handle, since they tend to stick to the fingers. There are two types of charges, positive and negative, with objects with like charges repelling each other and objects with unlike charges attracting each other. By the way, the acceleration of an object due to an electric force on it is still proportional to its mass. The amount of charge on the object may change the amount of force on it, but given a specific force the acceleration is strictly due to the mass.

The alert reader will notice that the definition above is for "electrical force" even though the topic under discussion is "electro*magnetic* forces". The electric force is straightforward: put a charged object in an electric field, it will be accelerated towards a concentration of opposite charges or away from a concentration of like charges. The magnetic field is harder to grasp.

If we have a permanent magnet in the shape of a ring with a gap cut out of it, there will be a magnetic field between the "magnetic poles" on either side of the gap. If we throw a charged object through the gap, it will not be accelerated towards the poles; it will instead be pushed out one side or other of the gap.

In more general terms, if a charged object flies across the lines of force of a magnetic field, the object is pushed *sideways* relative to its direction of motion, with a force proportional to the amount of charge on the object, the intensity of the magnetic field, and the velocity of the object.

It was recognized early on by researchers playing with electricity and magnetism that the two were related: a current passed through a wire will generate a magnetic field around it that can, for example, turn a compass needle.

In an inverse fashion, passing a wire through gap in a magnet would generate an electrical current. Further details of magnetism are a bit beyond the scope of this document and will need to be discussed elsewhere; however, it can be said without more explanation that magnetism is not a distinct force, actually being just a subtle manifestation of the same electrical force.

In general, an electric field is produced by an accumulation of charges, while a magnetic field is produced by *moving* charges. Why this was so couldn't be explained by classical physics, it was just accepted as "the rules".

Incidentally, Michael Faraday arrived at the concept of a field through the classic experiment of placing a magnet under a sheet of flat sheet of paper and then sprinkling iron filings on the paper; the filings then oriented themselves along the magnetic field, allowing it to be visualized.

It was a major step forward in the comprehension of the mysterious "action at a distance" of gravity and electromagnetism, though it really didn't say much about the underlying causes of either phenomenon. Faraday lacked a formal education, having dragged himself up from dire poverty at great effort, but he had one of the most well-developed senses of physical intuition in the history of the science.

As mentioned, electromagnetic forces are vastly more powerful than gravitational forces. If Dexter picks up two bricks and holds them out in parallel in front of him, he will not be able to sense any gravitational force between them, and in fact he wouldn't be able to sense any force with anything less than the most exacting experimental apparatus.

If he picks up two magnets, each the size of a brick, and holds them in front of him, they will slam together and he won't have the strength to pull them apart.

Both electromagnetism and gravity are long-range forces, and so it seems odd that gravity, which is so vastly weaker than electromagnetism, is the predominant force in the Universe at large scales.

There is a major difference between electromagnetism and gravity, however: as mentioned, electrical charges have "polarity" they can be positive or negative, with a repulsive force for like charges and an attractive force for unlike charges. In contrast, gravity is, as far as anyone has ever seen, is always attractive.

The strength of the electromagnetic force also limits its long-range effects. It is so strong that it tends to pull charges from one object to another. If Dexter scuffs his feet on a carpet to get a charge imbalance and then touches Deedee, charges jump between the two in a spark, correct the imbalance, and Deedee says: "Ouch!". Similarly, if on a larger scale, on a hot day rising air masses in thunderhead clouds will create a charge imbalance that will be corrected with a lightning bolt. Charge imbalances tend to be self-correcting, and overall the Universe remains electrically neutral.

Since gravity operates on a cosmic scale, the fact that it is very weak is highly significant: if it were substantially stronger, the entire Universe would collapse in on itself, and that would be the end of that. However, it must be emphasized that, except for gravity, the forces we see in our daily lives from our own physical work, the operation of a car or similar machine are derived from the electrical force, if sometimes very indirectly.

It should be noted that in recent years hints have arisen that there may be some sort of repulsive gravitational force. It was learned in the 20th century that the Universe is expanding, having been born in a highly dense and energetic state in a "Big Bang" billions of years ago, then expanding and cooling off to form planets, stars, and galaxies. The discovery that the Universe seems to be expanding at a faster and faster rate has suggested the existence of a mysterious "dark energy" that produces a negative gravity effect. However, nobody's really sure what dark energy is, and it's far beyond the scope of this document anyway.

As a footnote to fields, inventors of perpetual-motion machines and the like seem to have an odd attraction to fields. They particularly like to play with magnets. It doesn't take much inspection to show that fields violate neither energy conservation nor Newton's laws of motion.

As far as energy conservation goes, an object in a field can acquire no more energy than is put into it by work. The simplest example is raising a mass in a constant gravitation field. As described in the previous chapter, on Earth this requires an amount of work given by: 9.81 × height which is of course the potential energy acquired by the mass. The same reasoning applies to electric and magnetic fields, the only main difference being that it is much easier to manipulate or create these fields, for example by moving permanent magnets around or generating a magnetic field with a current. Moving the magnets and producing a current requires work, however, and there's no way to get more work out of any such arrangement than is put into it.

As far as Newton's laws go if Dexter drops an object on Earth, it is technically not as much true that it falls to the ground so much that the object and the Earth fall towards each other, each having the same but opposite momentum in the interaction. Since the Earth's mass is presumably vastly greater than that of the object, the velocity of the Earth is immeasurably smaller.

If it were a case of two planets of equal size, they would fall towards each other, accelerating at the same speed, and meet in the middle. The total momentum of the system at all times in any case is zero.

Another misunderstanding about fields is that they consume energy, even in static scenarios. This isn't true: although the Earth's gravity is pulling down on Wile E. Coyote's anvil at all times while it is hanging from a cliff, simply pulling on it doesn't use up any energy. The Coyote had to expend energy to get it to the top of the cliff, and it will release energy when it falls on him, but as long as it remains stationary, not the slightest amount of energy is expended.

Of course, what confuses the matter is that if the Coyote falls off the cliff and grabs on to a branch, he'll expend energy just hanging there. In reality, if a plastic dummy of the Coyote ballasted to have the same weight were tied to the branch, it would stay there indefinitely with no consumption of energy. The Coyote, like any living being, expends energy in biological processes for as long as he is alive and has to expend additional energy to perform any action, such as keeping his fingers wrapped around the tree branch.

From an energetic point of view, the Coyote is something like the kind of inflatable yard displays set up on holidays ghosts and giant spiders at Halloween, santas and snowmen at Christmas, monster bunnies at Easter. If they were airtight, they could be inflated and left in place with no additional input of work, but they aren't and their structure must be maintained by the continuous operation (and energy drain) of a small fan. If the energy supply to the fan is cut off, the display collapses, and if the Coyote doesn't have periodic inputs of energy, he collapses as well. This is why he is so desperate to catch that Roadrunner.

There is something mysterious about fields, and since they seem somewhat magical it is tempting to think that they are even more magical than they are. Despite that, from a fundamental physics point of view a gravitational field pulling together two masses is not all that different from, say, a stretched spring pulling together two masses. It's just that with a gravitational field the "spring" is invisible.

QUANTUM MECHANICS

Quantum Mechanics (QM) is a relatively new ground of physics that was developed at the beginning of the last century as the common effort of different scientists. Up to that time, it was possible to describe all known physical phenomena with the laws of classical physics as either *particles* (classical mechanics, kinematics) or *waves* (optics, classical electro-magnetism). The detection of new phenomena such as *e.g.* the *photoelectric effect* has lead to the development of a new physics that has revolutionized the way we look at the world around us. Many insights from quantum mechanics are at primary sight counterintuitive (*i.e.* they differ substantially from what we would expect in a classical picture!) which makes this field one of the most fascinating areas of modern physics. Quantum mechanics is a branch of physics dealing with physical phenomena where the action is on the arrange of the Planck constant. Quantum mechanics departs from classical mechanics primarily at the *quantum realm* of atomic and subatomic length scales. QM provides a mathematical description of much of the dual *particle-like* and *wave-like* behaviour and interactions of energy and matter.

In advanced topics of quantum mechanics, some of these behaviours are macroscopic and only emerge at extreme (*i.e.*, very low or very high) energies or temperatures. The name *quantum mechanics* derives from the observation that some physical quantities can change only in *discrete* amounts (Latin *quanta*), and not in a continuous (*cf.* analog) way. For example, the angular momentum of an electron bound to an atom or molecule is quantized. In the context of quantum mechanics, the wave-particle duality of energy and matter and the uncertainty principle provide a unified view of the behaviour of photons, electrons, and other atomic-scale objects.

The mathematical formulations of quantum mechanics are abstract. A mathematical purpose called the wavefunction provides information about the probability amplitude of position, momentum, and other physical properties of a particle. Mathematical manipulations of the wavefunction usually involve the braket notation, which requires an understanding of complex numbers and linear functionals.

The wavefunction treats the object as a quantum harmonic oscillator, and the mathematics is akin to that describing acoustic resonance. Many of the results of quantum mechanics are not easily visualized in terms of classical mechanics - for instance, the ground state in a quantum mechanical model is a non-zero energy state that is the lowest permitted energy state of a system, as opposed a more "traditional" system that is thought of as simply being at rest, with zero kinetic energy. Instead of a traditional static, unchanging zero state, quantum mechanics allows for far more dynamic, chaotic possibilities, according to John Wheeler.

The earliest versions of quantum mechanics were formulated in the first decade of the 20th century. At around the same time, the atomic theory and the corpuscular theory of light (as updated by Einstein) first came to be widely accepted as scientific fact; these latter theories can be viewed as quantum theories of matter and electromagnetic radiation, respectively. Early quantum theory was significantly reformulated in the mid-1920s by Werner Heisenberg, Max Born, Wolfgang Pauli and their collaborators, and the Copenhagen interpretation of Niels Bohr became widely accepted. By 1930, quantum mechanics had been further unified and formalized by the work of Paul Dirac and John von Neumann, with a greater emphasis placed on measurement in quantum mechanics, the statistical nature of our knowledge of reality, and philosophical speculation about the role of the observer.

Quantum mechanics has since branched out into almost every aspect of 20th century physics and other disciplines, such as quantum chemistry, quantum electronics, quantum optics, and quantum information science. Much 19th century physics has been re-evaluated as the "classical limit" of quantum mechanics, and its more advanced developments in terms of quantum field theory, string theory, and speculative quantum gravity theories.

PARTICLES AND WAVES

All phenomena in classical physics can be described either as a particle or a wave.

Wave-functions and Operators

The theory of quantum mechanics is built upon the fundamental concepts of *wave-functions* and *operators*. The wave-function is a single-valued square-integrable function of the system parameters and time which provides a complete description of the system. Linear Hermitian operators act on the wave-function and correspond to the physical *observables*, those dynamical variables which can be measured, *e.g.* position, momentum and energy. For systems of atomic nuclei and electrons, which are the subject of this dissertation, the system parameters might be taken to be a set of position variables of the constituent particles *i.e.*, $\{(r_i),(r_\alpha)\}$ their momenta $\{(p_i), (p_\alpha)$ or even a mixture of the two *e.g.* $\{(r_i),(p_\alpha)\}$. In contrast to a Newtonian system which is completely described by the positions *and* momenta of its constituents, the quantum-mechanical wave-function is a function of only one of these parameters per particle. The wave-function for the system is thus typically denoted by $\psi\{(r_i),(r_\alpha),t\}$.

A notation due to Dirac is often employed, which reflects the fact that this wave-function is simply one of many representations of a single *state-vector* in a Hilbert space, which is written as $|\Psi\rangle$, known as a *ket*.

There also exists a *dual space* containing a set of *bra* vectors, denoted $|\Psi\rangle$, defined by their scalar products and in one-to-one correspondence with the kets.

The scalar product is written as a *braket* and is anti-linear in the first argument and linear in the second: thus $|\Psi | \Phi\rangle = (\langle\Psi | \Phi\rangle)^*$.

It is worth noting here that state-vectors which differ only by a multiplicative non-zero complex constant describe the same state: we can thus restrict our interest to the set of *normalised* vectors defined such that the scalar product of the vector with its own conjugate equals unity:

$$\langle\Psi | \Psi\rangle = \int\prod_j dr_\beta \Psi^*\left(\{r_i\},\{r_a\},t\right) = 1.$$

The operator corresponding to some observable O is often written $\hat{o}$, and in general when this operator acts on some state-vector $|\Psi\rangle$, a different (not necessarily normalised) state-vector $|\Phi\rangle$ results:

$$\hat{O} | \Psi\rangle = | \Phi\rangle.$$

However, for each operator there exists a set of normalised *eigenstates*, say $\{|\chi_n\rangle\}$, which remain unchanged by the action of the operator *i.e.*

$$\hat{O} | \chi_n\rangle = \lambda_n | \chi_n\rangle$$

in which the constant λ_n (always real for Hermitian operators) is the *eigenvalue*.

The postulates of quantum mechanics state that for a system in state $|\Psi\rangle$:

- the outcome of a measurement of a dynamical variable is always one of the eigenvalues λ_n of the corresponding operator,
- immediately following a measurement, the state-vector collapses to the eigenstate $|\chi_n\rangle$ corresponding to the measured eigenvalue,
- the probability of such a measurement is

$$P(\lambda_n) = |\langle\chi_n | \Psi\rangle|^2.$$

Wave Phenomena

Not all phenomena of classical physics are describable in a particle picture. A second large class of physical events can be described as *waves*. A *wave* is a disturbance that propagates through space or space and time, often transferring energy.

From every day live, you are already familiar with many wave phenomena, such as for instance water waves or sound waves (periodical compressions of air that travel through space and can be detected by a mechanosensitive device in your inner ear).

Water waves and sound waves are examples for mechanical waves that exist in a medium (which on deformation is capable of producing elastic restoring forces).

Wave Phenomena: Interference and Diffraction

Waves can exhibit phenomena that cannot be explain in a classical particle pictures. Typical such wave phenomena are interference and diffraction. Two or more waves can superimpose and form a new wave, this effect is called interference. When two sinusoidal waves place over, the resulting waveform depends on the frequency, the amplitude and in particular on the relative difference in phase (the *phase shift*) of the two waves.

If the two waves have the same amplitude A and wavelength the resultant waveform will have an amplitude between 0 and $2A$ depending on whether the two waves are *in phase* (*i.e.* both waves have their minima and maxima at exactly the same time) or *out of phase* (*i.e.* one wave has a maximum when the other one has a minimum). Consider two waves that are in phase, with amplitudes A_1 and A_2. Their troughs and peaks line up and the resultant wave will have an amplitude $A = A_1 + A_2$. This is known as *constructive interference*.

If the two waves are π radians, or 180°, out of phase, then one wave's crests will coincide with another wave's troughs and so will tend to cancel out. The resultant amplitude is $A = |A1 - A2|$. If $A1 = A2$, the resultant amplitude will be zero. This is known as *destructive interference*.

RELATIVITY AND QUANTUM MECHANICS

Even with the defining postulates of both Einstein's theory of general relativity and quantum theory being indisputably supported by rigorous and repeated empirical evidence and while they do not directly contradict each other theoretically (at least with regard to primary claims), they are resistant to being incorporated within one cohesive model.

Einstein himself is well known for rejecting some of the claims of quantum mechanics. While clearly contributing to the field, he did not accept the more philosophical consequences and interpretations of quantum mechanics, such as the lack of deterministic causality and the assertion that a single subatomic particle can occupy numerous areas of space at one time. He also was the first to notice some of the apparently exotic consequences of entanglement and used them to formulate the Einstein-Podolsky-Rosen paradox, in the hope of showing that quantum mechanics had unacceptable implications. This was 1935, but in 1964 it was shown by John Bell (see Bell inequality) that, although Einstein was correct in identifying seemingly paradoxical implications of quantum mechanical non-locality, these implications could be experimentally tested. Alain Aspect's initial experiments in 1982, and many subsequent experiments since,, have verified quantum entanglement.

According to the paper of J. Bell and the Copenhagen interpretation (the common interpretation of quantum mechanics by physicists since,, 1927), and contrary to Einstein's ideas, quantum mechanics was not at the same time

- A *realistic* theory
- A *local* theory.

The Einstein-Podolsky-Rosen paradox shows in any case that there exist experiments by which one can measure the state of one particle and instantaneously change the state of its entangled partner, although the two particles can be an arbitrary distance apart; however, this effect does not violate causality, since,, no transfer of information happens. Quantum entanglement is at the basis of quantum cryptography, with high-security commercial applications in banking and government. Gravity is negligible in many areas of particle physics, so that unification between general relativity and quantum mechanics is not an urgent issue in those applications. However, the lack of a correct theory of quantum gravity is an important issue in cosmology and physicists' search for an elegant "theory of everything".

Thus, resolving the inconsistencies between both theories has been a major goal of twentieth- and twenty-first-century physics. Many prominent physicists, including Stephen Hawking, have laboured in the attempt to discover a theory underlying *everything*, combining not only different models of subatomic physics, but also deriving the universe's four forces —the strong force, electromagnetism, weak force, and gravity— from a single force or phenomenon.

While Stephen Hawking was initially a believer in the Theory of Everything, after considering Gödel's Incompleteness Theorem, concluded that one was not obtainable, and stated such publicly in his lecture, "Gödel and the end of physics" in 2002. One of the leaders in this field is Edward Witten, a theoretical physicist who formulated the groundbreaking M-theory, which is an attempt at describing the supersymmetrical based string theory.

Attempts at a Unified Field Theory

As of 2011, the quest to unify the fundamental forces through quantum mechanics is still ongoing. Quantum electrodynamics (or "quantum electromagnetism"), which is currently (in the perturbative regime at least) the most accurately tested physical theory, has been successfully merged with the weak nuclear force into the electroweak force and work is currently being done to merge the electroweak and strong force into the electrostrong force.

Current predictions state that at around 10^{14} GeV the three aforementioned forces are fused into a single unified field, Beyond this "grand unification," it is speculated that it may be possible to merge gravity with the other three gauge symmetries, expected to occur at roughly 10^{19} GeV. However—and while special relativity is parsimoniously incorporated into quantum electrodynamics—the expanded general relativity, currently the best theory describing the gravitation force, has not been fully incorporated into quantum theory.

PHILOSOPHICAL IMPLICATIONS

Since,, its inception, the many counter-intuitive results of quantum mechanics have provoked strong philosophical debate and many interpretations. Even fundamental issues such as Max Born's basic rules concerning probability amplitudes and probability distributions took decades to be appreciated by the society and leading scientists. Richard Feynman said, "I think I can safely say that nobody understands quantum mechanics."

The Copenhagen interpretation, due largely to the Danish theoretical physicist Niels Bohr, is the interpretation of the quantum mechanical formalism most widely accepted amongst physicists. According to it, the probabilistic nature of quantum mechanics is not a temporary feature which will eventually be replaced by a deterministic theory, but instead must be considered to be a final renunciation of the classical ideal of causality. In this interpretation, it is believed that any well-defined application of the quantum mechanical formalism must always make reference to the experimental arrangement, due to the complementarity nature of evidence obtained under different experimental situations.

Albert Einstein, himself one of the founders of quantum theory, disliked this loss of determinism in measurement. (A view paraphrased as "God does not play dice with the universe.") Einstein held that there should be a local hidden variable theory underlying quantum mechanics and that, consequently, the present theory was incomplete.

He produced a series of objections to the theory, the most famous of which has become known as the Einstein-Podolsky-Rosen paradox. John Bell showed that the EPR paradox led to experimentally testable differences between quantum mechanics and local realistic theories.

Experiments have been performed confirming the accuracy of quantum mechanics, thus demonstrating that the physical world cannot be described by local realistic theories. The Bohr-Einstein debates provide a vibrant critique of the Copenhagen Interpretation from an epistemological point of view.

The Everett many-worlds interpretation, formulated in 1956, holds that all the possibilities described by quantum theory simultaneously occur in a multiverse composed of mostly independent parallel universes.

This is not accomplished by introducing some new axiom to quantum mechanics, but on the contrary by removing the axiom of the collapse of the wave packet: All the possible consistent states of the measured system and the measuring apparatus (including the observer) are present in a real physical (not just formally mathematical, as in other interpretations) quantum superposition.

Such a superposition of consistent state combinations of different systems is called an entangled state. While the multiverse is deterministic, we perceive non-deterministic behaviour governed by probabilities, because we can observe only the universe, *i.e.* the consistent state contribution to the mentioned superposition, we inhabit.

Everett's interpretation is perfectly consistent with John Bell's experiments and makes them intuitively understandable.

However, according to the theory of quantum decoherence, the parallel universes will never be accessible to us. This inaccessibility can be understood as follows: Once a measurement is done, the measured system becomes entangled with both the physicist who measured it and a huge number of other particles, some of which are photons flying away towards the other end of the universe; in order to prove that the wave function did not collapse one would have to bring all these particles back and measure them again, together with the system that was measured originally. This is completely impractical, but even if one could theoretically do this, it would destroy any evidence that the original measurement took placc (including the physicist's memory).

APPLICATIONS

Quantum mechanics had enormous success in explaining many of the features of our world. The individual behaviour of the subatomic particles that make up all forms of matter—electrons, protons, neutrons, photons and others—can often only be satisfactorily described using quantum mechanics. Quantum mechanics has strongly influenced the string theory, a candidate for a theory of everything and the multiverse hypothesis.

Quantum mechanics is important for understanding how individual atoms combine covalently to form chemicals or molecules. The application of quantum mechanics to chemistry is known as quantum chemistry. (Relativistic) quantum mechanics can in principle mathematically describe most of chemistry.

Quantum mechanics can provide quantitative insight into ionic and covalent bonding processes by explicitly showing which molecules are energetically favourable to which others, and by approximately how much. Most of the calculations performed in computational chemistry rely on quantum mechanics.

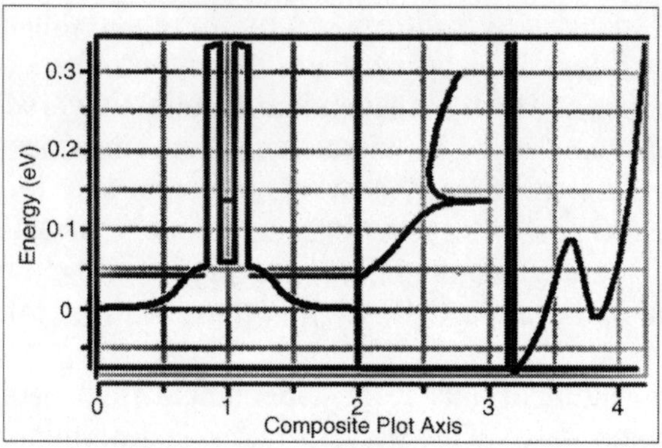

Fig. A Working Mechanism of a Resonant Tunneling Diode Device, Based on the Phenomenon of Quantum Tunneling Through the Potential Barriers.

Much of modern technology operates at a scale where quantum effects are significant. Examples include the laser, the transistor (and thus the microchip), the electron microscope, and magnetic resonance imaging. The study of semiconductors led to the invention of the diode and the transistor, which are indispensable for modern electronics. Researchers are currently seeking robust methods of directly manipulating quantum states. Efforts are being made to develop quantum cryptography, which will allow guaranteed secure transmission of information.

A more distant goal is the development of quantum computers, which are expected to perform certain computational tasks exponentially faster than classical computers. Another active research topic is quantum teleportation, which deals with techniques to transmit quantum information over arbitrary distances.

Quantum tunneling is vital in many devices, even in the simple light switch, as otherwise the electrons in the electric current could not penetrate the potential barrier made up of a layer of oxide. Flash memory chips found in USB drives use quantum tunneling to erase their memory cells.

Quantum mechanics primarily applies to the atomic regimes of matter and energy, but some systems exhibit quantum mechanical effects on a large scale; superfluidity (the frictionless flow of a liquid at temperatures near absolute zero) is one well-known example. Quantum theory also provides accurate descriptions for many previously unexplained phenomena such as black body radiation and the stability of electron orbitals. It has also given insight into the workings of many different biological systems, including smell receptors and protein structures.

Recent work on photosynthesis has provided evidence that quantum correlations play an essential role in this most fundamental process of the plant kingdom. Even so, classical physics often can be a good approximation to results otherwise obtained by quantum physics, typically in circumstances with large numbers of particles or large quantum numbers. (However, some open questions remain in the field of quantum chaos.)

PHYSICAL PROPERTIES—DENSITY, VISCOSITY, AND SURFACE TENSION

There are three physical properties of fluids that are particularly important: density, viscosity, and surface tension. Each of these will be defined and viewed briefly in terms of molecular concepts, and their dimensions will be examined in terms of mass, length, and time (M, L, and T). The physical properties depend primarily on the particular fluid.

For liquids, viscosity also depends strongly on the temperature; for gases, viscosity is approximately proportional to the square root of the absolute temperature.

The density of gases depends almost directly on the absolute pressure; for most other cases, the effect of pressure on physical properties can be disregarded.

Typical processes often run almost isothermally, and in these cases the effect of temperature can be ignored. Except in certain special cases, such as the flow of a compressible gas (in which the density is not constant) or a liquid under a very high shear rate (in which viscous dissipation can cause significant internal heating), or situations involving exothermic or endothermic reactions, we shall ignore any variation of physical properties with pressure and temperature.

Densities of liquids. Density depends on the mass of an individual molecule and the number of such molecules that occupy a unit of volume. For liquids, density depends primarily on the particular liquid and, to a much smaller extent, on its temperature.

Representative densities of liquids are given in table. The accuracy of the values given in tables is adequate for the calculations needed in this text. However, if highly accurate values are needed, particularly at extreme conditions, then specialized information should be sought elsewhere.

Density: The density ρ of a fluid is defined as its mass per unit volume, and indicates its inertia or resistance to an accelerating force. Thus:

$$\rho = \frac{\text{mass}}{\text{volume}}[=]\frac{M}{L^3},$$

in which the notation "[=]" is consistently used to indicate the dimensions of a quantity. It is usually understood in Equation. that the volume is chosen so that it is neither so small that it has no chance of containing a representative selection of molecules nor so large that (in the case of gases) changes of pressure cause significant changes of density throughout the volume. A medium characterized by a density is called a continuum, and follows the classical laws of mechanics— including Newton's law of motion.

Table: Specific Gravities, Densities, and Thermal Expansion Coefficients of Liquids at 20 °C

Liquid	Sp. Gr. s	Density, kg/m^3	ρ lb$_m$/ft^3	α °C^{-1}
Acetone	0.792	792	49.4	0.00149
Benzene	0.879	879	54.9	0.00124
Crude oil, 35°API	0.851	851	53.1	0.00074
Ethanol	0.789	789	49.3	0.00112
Glycerol	1.26 (50 °C)	1,260	78.7	—
Kerosene	0.819	819	51.1	0.00093
Mercury	13.55	13,550	845.9	0.000182
Methanol	0.792	792	49.4	0.00120
n-Octane	0.703	703	43.9	—
n-Pentane	0.630	630	39.3	0.00161
Water	0.998	998	62.3	0.000207

Degrees A.P.I. (American Petroleum Institute) are related to specific gravity s by the formula:

$$°\text{A.P.I.} = \frac{141.5}{s} - 131.5$$

Note that for water, °A.P.I. = 10, with correspondingly higher values for liquids that are less dense. Thus, for the crude oil listed in Table, equation. indeed gives 141.5/0.851 – 131.5 = 35°A.P.I.

Densities of gases. For ideal gases, $pV = nRT$, where p is the absolute pressure, V is the volume of the gas, n is the number of moles (abbreviated as "mol" when used as a unit), R is the gas constant, and T is the absolute temperature. If M_w is the molecular weight of the gas, it follows that:

$$\rho = \frac{nM_w}{V} = \frac{M_w p}{RT}.$$

Thus, the density of an ideal gas depends on the molecular weight, absolute pres-sure, and absolute temperature. Values of the gas constant R are given in Table for various systems of units. Note that degrees Kelvin, formerly represented by " °K," is now more simply denoted as "K."

Table: Values of the Gas Constant, R

Value	Units
8.314	J/g-mol K
0.08314	litre bar/g-mol K
0.08206	litre atm/g-mol K
1.987	cal/g-mol K
10.73	psia ft^3 /lb-mol °R
0.7302	ft^3 atm/lb-mol °R
1,545	ft lb$_f$ /lb-mol °R

For a non-ideal gas, the compressibility factor Z (a function of p and T) is introduced into the denominator of equation, giving:

$$\rho = \frac{nM_w}{V} = \frac{M_w p}{ZRT}.$$

Thus, the extent to which Z deviates from unity gives a measure of the non-ideality of the gas.

The isothermal compressibility of a gas is defined as:

$$\beta = \frac{1}{V}\left(\frac{\partial V}{\partial p}\right)_T,$$

and equals—at constant temperature—the fractional decrease in volume caused by a unit increase in the pressure. For an ideal gas, $\beta = 1/p$, the reciprocal of the absolute pressure.

The coefficient of thermal expansion α of a material is its isobaric (constant pressure) fractional increase in volume per unit rise in temperature:

$$\alpha = \frac{1}{V}\left(\frac{\partial V}{\partial T}\right)_p,$$

Since, for a given mass, density is inversely proportional to volume, it follows that for moderate temperature ranges (over which α is essentially constant) the density of most liquids is approximately a linear function of temperature:

$$\rho = \rho_0[1 - \alpha(T - T_0)],$$

where ρ_0 is the density at a reference temperature T_0. For an ideal gas, ⟨ = 1/ T, the reciprocal of the absolute temperature.

The specific gravity s of a fluid is the ratio of the density ρ to the density ρ_{SC} of a reference fluid at some standard condition:

$$s = \frac{\rho}{\rho_{SC}}.$$

For liquids, ρ_{SC} is usually the density of water at 4 °C, which equals 1.000 g/ml or 1,000 kg/m^3. For gases, ρ_{SC} is sometimes taken as the density of air at 60 °F and 14.7 psia, which is approximately 0.0759 lb$_m$/ft^3, and sometimes at 0 °C and one atmosphere absolute; since there is no single standard for gases, care must obviously be taken when interpreting published values. For natural gas, consisting primarily of methane and other hydrocarbons, the gas gravity is defined as the ratio of the molecular weight of the gas to that of air (28.8 lb$_m$ / lb-mol).

Values of the molecular weight M_w are listed in Table for several commonly occurring gases, together with their densities at standard conditions of atmospheric pressure and 0 °C.

Table: Gas Molecular Weights and Densities (the Latter at Atmospheric Pressure and 0 °C)

Gas	Mw	Standard kg/m^3	Density lb$_m$/ft^3
Air	28.8	1.29	0.0802

Carbon dioxide	44.0	1.96	0.1225
Ethylene	28.0	1.25	0.0780
Hydrogen	2.0	0.089	0.0056
Methane	16.0	0.714	0.0446
Nitrogen	28.0	1.25	0.0780
Oxygen	32.0	1.43	0.0891

Viscosity: The viscosity of a fluid measures its resistance to flow under an applied shear stress. There, the fluid is ideally supposed to be confined in a relatively small gap of thickness h between one plate that is stationary and another plate that is moving steadily at a velocity V relative to the first plate.

In practice, the situation would essentially be realized by a fluid occupying the space between two concentric cylinders of large radii rotating relative to each other. A steady force F to the right is applied to the upper plate (and, to preserve equilibrium, to the left on the lower plate) in order to maintain a constant motion and to overcome the viscous friction caused by layers of molecules sliding over one another.

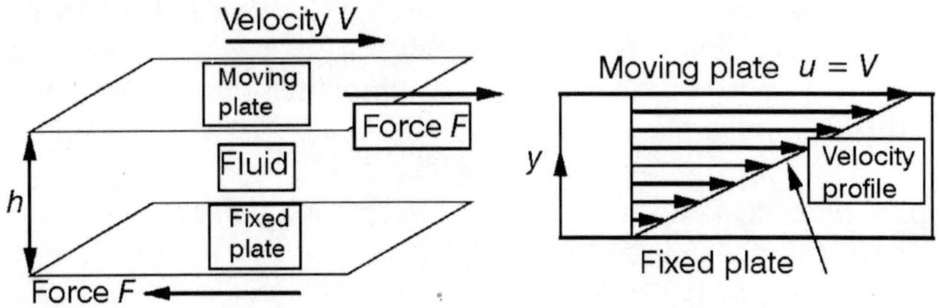

(a) (b)

Fig. (a) Fluid in shear between parallel plates;
(b) the ensuing linear velocity profile.

Under these circumstances, the velocity u of the fluid to the right is found experimentally to vary linearly from zero at the lower plate ($y = 0$) to V itself at the upper plate, corresponding to no-slip conditions at each plate. At any intermediate distance y from the lower plate, the velocity is simply:

$$u = \frac{y}{h}V.$$

Recall that the shear stress Ä is the tangential applied force F per unit area:

$$\tau = \frac{F}{A},$$

in which A is the area of each plate. Experimentally, for a large class of materials, called Newtonian fluids, the shear stress is directly proportional to the velocity gradient:

$$\tau = \mu\frac{du}{dy} = \mu\frac{V}{h}.$$

The proportionality constant μ is called the viscosity of the fluid; its dimensions can be found by substituting those for F (ML/T^2), A (L^2), and du/dy (T^{-1}), giving:

$$\mu \, [=] \, \frac{M}{LT}.$$

Representative units for viscosity are g/cm s (also known as poise, designated by P), kg/m s, and lb_m/ft hr. The centipoise (cP), one hundredth of a poise, is also a convenient unit, since the viscosity of water at room temperature is approximately 0.01 P or 1.0 cP. Table gives viscosity conversion factors.

The viscosity of a fluid may be determined by observing the pressure drop when it flows at a known rate in a tube. More sophisticated methods for determining the rheological or flow properties of fluids—including viscosity; such methods often involve containing the fluid in a small gap between two surfaces, moving one of the surfaces, and measuring the force needed to maintain the other surface stationary.

Table: Viscosity Parameters for Liquids

Liquid	*a*	*b*	*a*	*b*
	(*T* in K)		(*T* in °*R*)	
Acetone	14.64	–2.77	16.29	–2.77
Benzene	21.99	–3.95	24.34	–3.95
Crude oil, 35° API	53.73	–9.01	59.09	–9.01
Ethanol	31.63	–5.53	34.93	–5.53
Glycerol	106.76	–17.60	117.22	–17.60
Kerosene	33.41	–5.72	36.82	–5.72
Methanol	22.18	–3.99	24.56	–3.99
Octane	17.86	–3.25	19.80	–3.25
Pentane	13.46	–2.62	15.02	–2.62
Water	29.76	–5.24	32.88	–5.24

The kinematic viscosity ν is the ratio of the viscosity to the density:

$$\nu = \frac{\mu}{\rho},$$

and is important in cases in which significant viscous and gravitational forces coexist. The reader can check that the dimensions of ν are L^2/T, which are identical to those for the diffusion coefficient $\mathcal{D}$ in mass transfer and for the thermal diffusivity $\acute{a} = k/\rho c_p$ in heat transfer. There is a definite analogy among the three quantities—indeed, as seen later, the value of the kinematic viscosity governs the rate of "diffusion" of momentum in the laminar and turbulent flow of fluids.

Viscosities of liquids. The viscosities μ of liquids generally vary approximately with absolute temperature T according to:

$$\ln \mu = a + b \ln T \text{ or } \mu. = e^{a+b\ln T},$$

and—to a good approximation—are independent of pressure. Assuming that μ

is measured in centipoise and that T is either in degrees Kelvin or Rankine, appro-priate parameters a and b are given in Table for several representative liquids. The resulting values for viscosity are approximate, suitable for a first design only.

Viscosities of gases. The viscosity μ of many gases is approximated by the formula:

$$\mu = \mu_0 \left(\frac{T}{T_0} \right)^n,$$

in which T is the absolute temperature (Kelvin or Rankine), μ_0 is the viscosity at an absolute reference temperature T_0, and n is an empirical exponent that best fits the experimental data. The values of the parameters μ_0 and n for atmospheric pressure are given in Table ; recall that to a first approximation, the viscosity of a gas is independent of pressure. The values μ_0 are given in centipoise and correspond to a reference temperature of T_0.= 273 K.= 492 °R.

Table: Viscosity Parameters for Gases

Gas	μ_0, *cP*	*n*
Air	0.0171	0.768
Carbon dioxide	0.0137	0.935
Ethylene	0.0096	0.812
Hydrogen	0.0084	0.695
Methane	0.0120	0.873
Nitrogen	0.0166	0.756
Oxygen	0.0187	0.814

Surface tension: Surface tension is the tendency of the surface of a liquid to behave like a stretched elastic membrane. There is a natural tendency for liquids to minimize their surface area. The obvious case is that of a liquid droplet on a horizontal surface that is not wetted by the liquid—mercury on glass, or water on a surface that also has a thin oil film on it. For small droplets, the droplet adopts a shape that is almost perfectly spherical, because in this configuration there is the least surface area for a given volume.

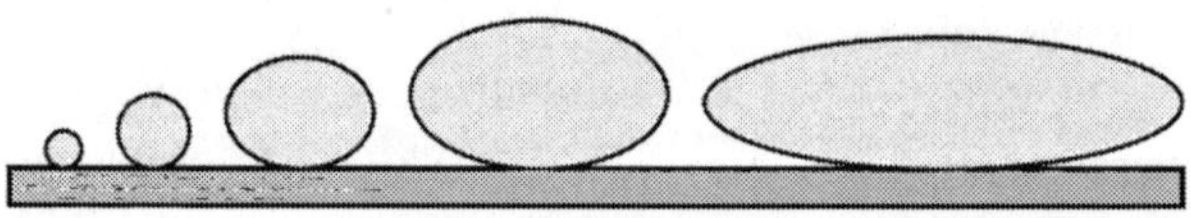

Fig. The larger droplets are flatter because gravity is becoming more important than surface tension.

For larger droplets, the shape becomes somewhat flatter because of the increasingly important gravitational effect, which is roughly proportional to a^3, where a is the approximate droplet radius, whereas the surface area is proportional only to a^2. Thus, the ratio of gravitational to surface tension effects depends roughly on the value of a^3 / a^2 = a, and is therefore increasingly important for the larger droplets. Overall, the situation is very similar to that of a water-filled balloon, in which the water accounts for the gravitational effect and the balloon acts like the surface tension.

A fundamental property is the surface energy. A molecule I, situated in the interior of the liquid, is attracted equally in all directions by its neighbors. However, a molecule S, situated in the surface, experiences a net attractive force into the bulk of the liquid. (The vapor above the surface, being comparatively rarefied, exerts a negligible force on molecule S.) Therefore, work has to be done against such a force in bringing an interior molecule to the surface. Hence, an energy σ, called the surface energy, can be attributed to a unit area of the surface.

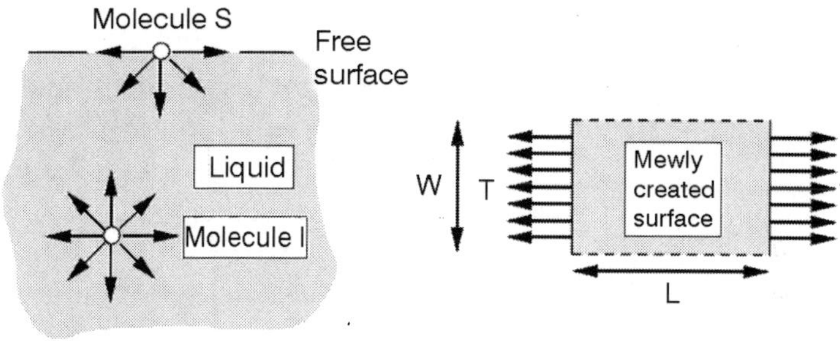

Fig. (a) Molecules in the interior and surface of a liquid; (b) newly created surface caused by moving the tension T through a distance L.

An equivalent viewpoint is to consider the surface tension T existing per unit distance of a line drawn in the surface. Suppose that such a tension has moved a distance L, thereby creating an area WL of fresh surface. The work done is the product of the force, TW, and the distance L through which it moves, namely TWL, and this must equal the newly acquired surface energy σWL. Therefore, $T = \sigma$; both quantities have units of force per unit distance, such as N/m, which is equivalent to energy per unit area, such as J/m^2.

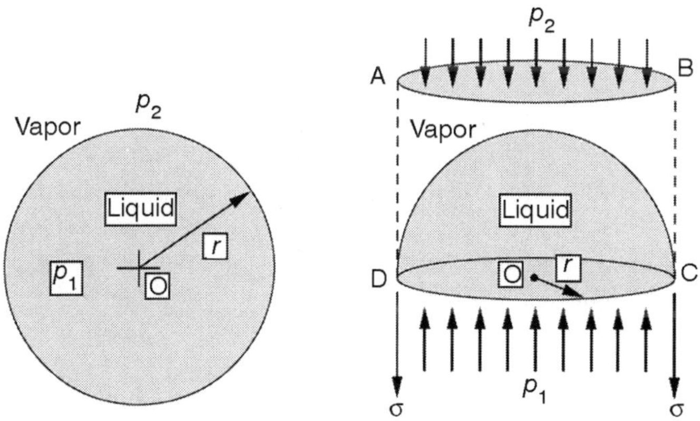

(a) Liquid droplet (b) Forces in equilibrium

Fig. Pressure change across a curved surface.

We next find the amount $p_1 - p_2$ by which the pressure p_1 inside a liquid droplet of radius r, exceeds the pressure p_2 of the surrounding vapor. The equilibrium of the upper hemisphere of the droplet, which is also surrounded by an imaginary cylindrical "control surface" ABCD, on which forces in the vertical direction will soon be equated. Observe that the internal pressure p_1 is trying to blow apart the two hemispheres (the lower one is not shown), whereas the surface tension σ is trying to pull them together.

In more detail, there are two different types of forces to be considered:

- That due to the pressure diverence between the pressure inside the droplet and the vapor outside, each acting on an area πr^2 (that of the circles CD and AB):
$$(p_1 - p_2)\pi r^2.$$
- That due to surface tension, which acts on the circumference of length $2\pi r$:
$$2\pi r\sigma.$$

At equilibrium, these two forces are equated, giving:

$$\Delta p = p_1 - p_2 =$$

That is, there is a higher pressure on the concave or droplet side of the interface.

What would the pressure change be for a bubble instead of a droplet? Why? More generally, if an interface has principal radii of curvature r1 and r2, the increase in pressure can be shown to be:

$$p_1 - p_2 = \left(\frac{1}{r_1} + \frac{1}{r_2}\right)$$

For a sphere of radius r, both radii are equal, so that $r_1 = r_2 = r$, and $p_1 - p_2 = 2\sigma/r$. The radii r_1 and r_2 will have the same sign if the corresponding centers of curvature are on the same side of the interface; if not, they will be of opposite sign. Appendix A contains further information about the curvature of a surface.

A brief description of simple experiments for measuring the surface tension σ of a liquid:

- In the *capillary-rise* method, a narrow tube of internal radius a is dipped vertically into a pool of liquid, which then rises to a height h inside the tube; if the *contact angle* (the angle between the free surface and the wall) is è, the meniscus will be approximated by part of the surface of a sphere; from the geometry shown in the enlargement on the right-hand side of Fig the radius of the sphere is seen to be $r = a/\cos$ è. Since the surface is now concave on the air side, the reverse of equation occurs, and $p_2 = p_1 - 2ó/r$, so that p_2 is *below* atmospheric pressure p_1. Now follow the path 1–2–3–4, and observe that $p_4 = p_3$ because points 3 and 4 are at the same elevation in the same liquid. Thus, the pressure at point 4 is:

$$p_4 = p_1 - \frac{2\sigma}{r} + \rho g h.$$

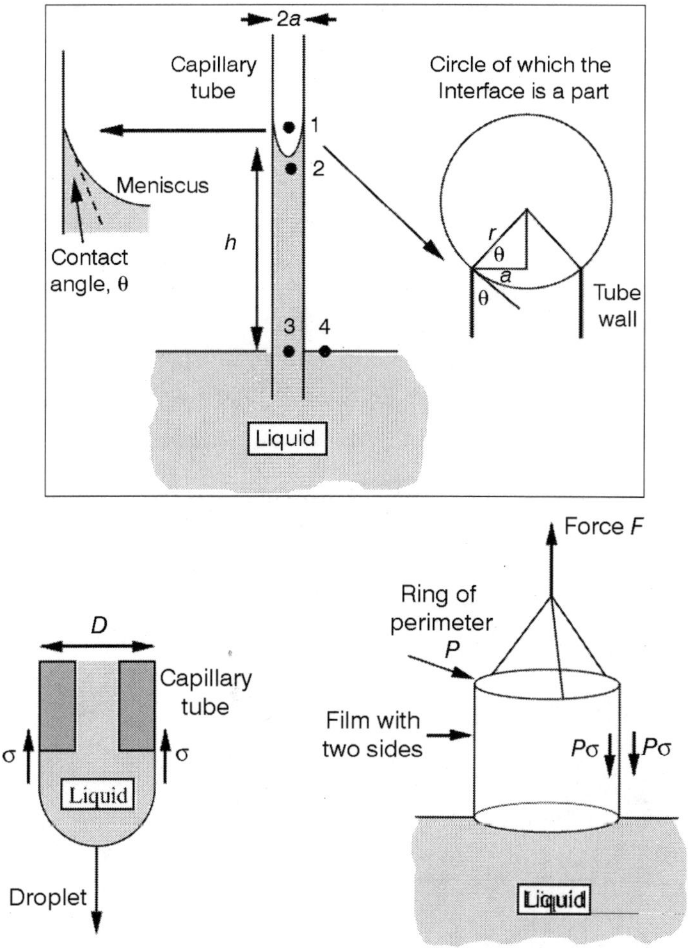

Fig. Methods for measuring surface tension.

However, $p_4 = p_1$ since both of these are at atmospheric pressure. Hence, the surface tension is given by the relation:

$$\sigma = \frac{1}{2}\rho g h r = \frac{\rho g h a}{2\cos\theta}.$$

In many cases—for complete wetting of the surface—*è* is essentially zero and cos *è* = 1. However, for liquids such as mercury in glass, there may be a complete *non*-wetting of the surface, in which case *è* = *ð*, so that cos *è* = – 1; the result is that the liquid level in the capillary is then *depressed* below that in the surrounding pool.

- In the *drop-weight* method, a liquid droplet is allowed to form very slowly at the tip of a capillary tube of outer diameter D. The droplet will eventually grow to a size where its weight just overcomes the surface-tension force ðDó holding it up. At this stage, it will detach from the tube, and its weight $w = Mg$ can be determined by catching it in a small pan and weighing it. By equating the two forces, the surface tension is then calculated from:

$$\sigma = \frac{w}{\pi D}$$

- In the *ring tensiometer,* a thin wire ring, suspended from the arm of a sensitive balance, is dipped into the liquid and gently raised, so that it brings a thin liquid film up with it. The force F needed to support the film is measured

by the balance. The downward force exerted on a unit length of the ring by one side of the film is the surface tension; since there are two sides to the film, the total force is 2*P*ó, where P is the circumference of the ring. The surface tension is therefore determined as:

$$\sigma = \frac{F}{2P}.$$

In common with most experimental techniques, all three methods described above require slight modifications to the results expressed in equations because of imperfections in the simple theories.

Surface tension generally appears only in situations involving either free surfaces (liquid/gas or liquid/solid boundaries) or interfaces (liquid/liquid boundaries); in the latter case, it is usually called the *interfacial* tension. Representative values for the surface tensions of liquids at 20°C, in contact either with air or their vapor (there is usually little difference between the two),are given in Table.

Table: Surface Tensions

Liquid	ó dynes/cm
Acetone	23.70
Benzene	28.85
Ethanol	22.75
Glycerol	63.40
Mercury	435.5
Methanol	22.61
n-Octane	21.80
Water	72.75

UNITS AND SYSTEMS OF UNITS

MASS, WEIGHT, AND FORCE.

The *mass M* of an object is a measure of the amount of matter it contains and will be constant, since it depends on the number of constituent molecules and their masses.

On the other hand, the *weight w* of the object is the gravitational force on it, and is equal to *Mg*, where *g* is the local gravitational acceleration. Mostly, we shall be discussing phenomena occurring at the surface of the earth, where *g* is approximately 32.174 ft/s^2 = 9.807 m/s^2 = 980.7 cm/s^2. These values are simply taken as 32.2, 9.81, and 981, respectively.

Table: Representative Units of Force

System	Units of Force	Customary Name
SI	kg m/s^2	Newton
CGS	g cm/s^2	Dyne
FPS	lbm ft/s^2	Poundal

Newton's second law of motion states that a force *F* applied to a mass *M* will give it an acceleration *a*:

$$F = ma,$$

From which is apparent that force has *dimensions* ML/T^2. Table gives the corresponding *units* of force in the SI (meter/kilogram/second), CGS (centimeter/ gram/second), and FPS (foot/pound/second) systems.

The poundal is now an archaic unit, hardly ever used. Instead, the *pound force*, lb_f , is much more common in the English system; it is defined as the gravitational force on 1 lb_m, which, if left to fall freely, will do so with an acceleration of 32.2 ft/s^2. Hence:

$$1\ lb_f = 32.2 lb_m \frac{ft}{s^2} = 32.2 \text{ poundals.}$$

When using lb_f in the ft, lb_m, s (FPS) system, the following conversion factor, commonly called "g_c," will almost invariably be needed:

$$g_c = 32.2 \frac{lb_m ft/s^2}{lb_f} = 32.2 \frac{lb_m ft}{lb_f s^2}.$$

Some writers incorporate g_c into their equations, but this approach may be confusing since it virtually implies that one particular set of units is being used, and hence tends to rob the equations of their generality.

Why not, for example, also incorporate the conversion factor of 144 in^2/ft^2 into equations where pressure is expressed in lbf/in$_2$? We prefer to omit all conversion factors in equations, and introduce them only as needed in evaluating expressions numerically.

If the reader is in any doubt, units should *always* be checked when performing calculations.

SI Units.

The most systematically developed and universally accepted set of units occurs in the *SI* units or *Systeme International d'Unit´es*; the subset we mainly need is shown in Table.

Table: SI Units

Physical Quantity	*Name of Unit*	*Symbol for Unit*	*Definition of Unit*
Basic Units			
Length	meter	m	–
Mass	kilogram	kg	–
Time	second	s	–
Temperature	degree Kelvin	K	–
Supplementary Unit			
Plane angle	radian	rad	—
Derived Units			
Acceleration			m/s^2
Angular velocity			rad/s
Density			kg/m^3
Energy	joule	J	kg m^2/s^2
Force	newton	N	kg m/s^2
Kinematic viscosity			m^2/s
Power	watt	W	kg m^2/s^3 (J/s)
Pressure	pascal	Pa	kg/m s^2 (N/m2)
Velocity			m/s
Viscosity			kg/m s

The basic units are again the meter, kilogram, and second (m, kg, and s); from these, certain *derived* units can also be obtained. Force (kg m/s^2) has already been discussed; energy is the product of force and length; power amounts to energy per unit time; surface tension is energy per unit area or force per unit length, and so on. Some of the units have names, and these, together with their abbreviations.

Tradition dies hard, and certain other "metric" units are so well established that they may be used as *auxiliary* units; these are shown in Table. The *gram* is the classic example. Note that the basic SI unit of mass (kg) is even represented in terms of the gram, and has not yet been given a name of its own.

Table: Auxiliary Units Allowed in Conjunction with SI Units

Physical Quantity	*Name of Unit for*	*Symbol Unit*	*Definition of Unit*
Area	hectare	ha	10^4 m^2
Kinematic viscosity	stokes	St	10^{-4} m^2/s
Length	micron	ìm	10^{-6} m
Mass	tonne	t	10^3 kg = Mg
	gram	g	10^{-3} kg = g
Pressure	bar	bar	10^5 N/m^2
Viscosity	poise	P	10^{-1} kg/m s
Volume	litre	l	10^{-3} m^3

Table shows some of the acceptable prefixes that can be used for accommodating both small and large quantities. For example, to avoid an excessive number of decimal places, 0.000001 s is normally better expressed as 1 ìs (one microsecond). Note also, for example, that 1 ìkg should be written as 1 mg—one prefix being better than two.

Table: Prefixes for Fractions and Multiples

Factor	*Name*	*Symbol*	*Factor*	*Name*	*Symbol*
10^{-12}	pico	p	10^{3}	kilo	k
10^{-9}	nano	n	10^{6}	mega	M
10^{-6}	micro	ì	10^{9}	giga	G
10^{-3}	milli	m	10^{12}	tera	T

Some of the more frequently used conversion factors are given in Table.

Table: Commonly Used Conversion Factors

Area	1 mile^2	=	640 acres
	1 acre	=	0.4047 ha
Energy	1 BTU	=	1,055 J
	1 cal	=	4.184 J
	1 J	=	0.7376 ft lbf
	1 erg	=	1 dyne cm
Force	1 lb_f	=	4.448 N
	1 N	=	0.2248 lbf
Length	1 ft	=	0.3048 m
	1 m	=	3.281 ft
	1 mile	=	5,280 ft
Mass	1 lbm	=	0.4536 kg
	1 kg	=	2.205 lb_m
Power	1 HP	=	550 ft lb_f/s
	1 kW	=	737.6 ft lb_f/s
Pressure	1 atm	=	14.696 lb_f/in^2
	1 atm	=	1.0133 bar
	1 atm	=	1.0133 × 105 Pa
Time	1 day	=	24 hr
	1 hr	=	60 min
	1 min	=	60 s
Viscosity	1 cP	=	2.419 lb_m/ft hr
	1 cP	=	0.001 kg/ms
	1 cP	=	0.000672 lb_m/ft s
	1 lb_f s/ft^2	=	4.788 × 10^4 cP
Volume	1 ft^3	=	7.481 U.S. gal
	1 U.S. gal	=	3.785 l
	1 m^3	=	264.2 U.S. gal

4

Hydrostatic Force on a Curved Surface

A submarine, whose hull has a circular cross section of diameter D, is just submerged in water of density $\tilde{n}$. Derive an equation that gives the total horizontal force Fx on the left half of the hull, for a distance W normal to the plane of the diagram. If D = 8 m, the circular cross section continues essentially for the total length W = 50 m of the submarine, and the density of sea water is $\tilde{n}$ = 1,026 kg/m_3, determine the total horizontal force on the left-hand half of the hull.

Solution: The force is obtained by evaluating the integral of equation, which is identical to that for the rectangle:

$$F_x = \int_{A*} p\,dA = \int_{z=0}^{z=D} \rho g W(D-z)dz = \frac{1}{2}\rho g W D^2.$$

Insertion of the numerical values gives:

$$F_x = \frac{1}{2} \times 1{,}026 \times 9.81 \times 50 \times 8.0^2 = 1.61 \times 10^7\,N.$$

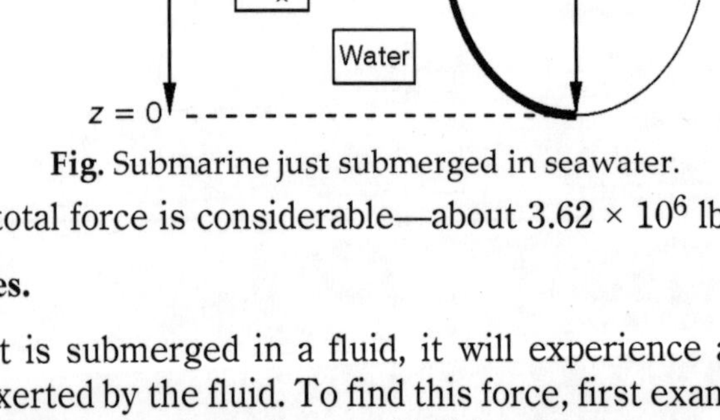

Fig. Submarine just submerged in seawater.

Thus, the total force is considerable—about 3.62 × 10^6 lb_f.

Buoyancy forces.

If an object is submerged in a fluid, it will experience a net upward or *buoyant* force exerted by the fluid. To find this force, first examine the buoyant force on a submerged circular cylinder of height H and cross-sectional area A.

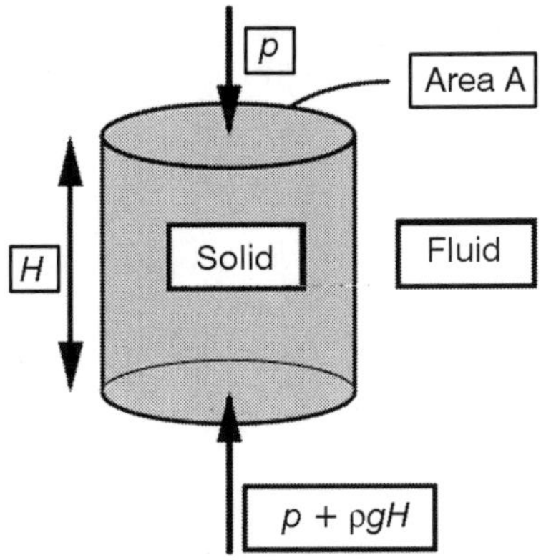

Fig. Pressure forces on a submerged cylinder.

The forces on the curved vertical surface act horizontally and may therefore be ignored. Hence, the net *upward* force due to the difference between the opposing pressures on the bottom and top faces is:

$F = (p + \rho g H - p)A = \rho H A g$, which is exactly the weight of the displaced liquid, thus verifying *Archimedes' law*, (the buoyant force equals the weight of the fluid displaced) for the cylinder. The same result would clearly be obtained for a cylinder of any uniform cross section.

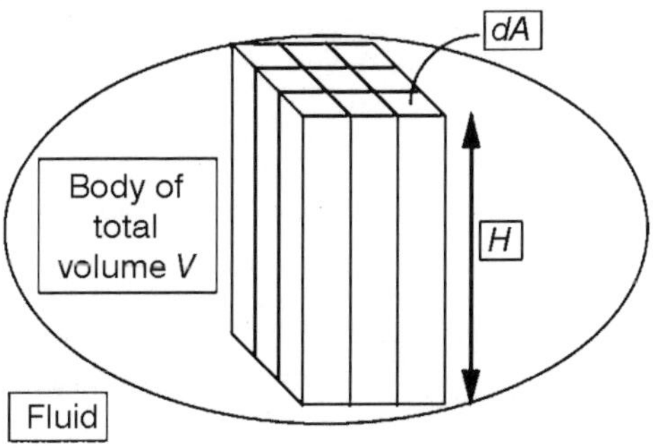

Fig. Buoyancy force for an arbitrary shape.

Figure shows a more general situation, with a body of arbitrary shape. However, Archimedes' law still holds since the body can be decomposed into an infinitely large number of vertical rectangular parallelepipeds or "boxes" of infinitesimally small cross-sectional area *dA*. The effect for one box is then summed or "integrated" over all the boxes, and again gives the net upward buoyant force as the weight of the liquid displaced.

APPLICATION OF ARCHIMEDES' LAW

Consider the situation in figure, in which a barrel rests on a raft that floats in a swimming pool. The barrel is then pushed off the raft, and may either float or sink, depending on its contents and hence its mass. The cross-hatching shows the volumes of water that are displaced. For each of the cases, determine whether the water level in the pool will rise, fall, or remain constant, relative to the initial level in (a).

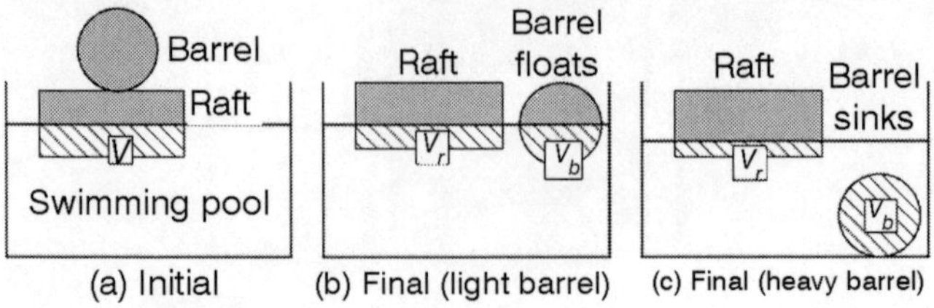

(a) Initial (b) Final (light barrel) (c) Final (heavy barrel)

Fig. Raft and barrel in swimming pool: (a) initial positions, (b) light barrel rolls off and floats, (c) heavy barrel rolls off and sinks. The cross-hatching shows volumes below the surface of the water.

Solution: *Initial state.* Let the masses of the raft and barrel be *Mr* and *Mb*, respectively. If the volume of displaced water is initially *V* in (a), Archimedes' law requires that the total weight of the raft and barrel equals the weight of the displaced water, whose density is

$$\rho : (M_r + M_b)g = V\rho g.$$

Barrel floats. If the barrel floats, as in (b), with submerged volumes of *Vr* and *Vb* for the raft and barrel, respectively, Archimedes' law may be applied to the raft and barrel separately:

$$\text{Raft: } m_r g = V_r \rho g, \text{ Barrel: } m_b g = V_b \rho g.$$

Addition and compare of the two equations shows that:

$$V_r + V_b = V.$$

Therefore, since the volume of the water is constant, and the total displaced volume does not change, the level of the surface also remains *unchanged*. *Barrel sinks.* Archimedes' law may still be applied to the raft, but the weight of the water displaced by the barrel no longer suffices to support the weight of the barrel, so that:

$$\text{Raft: } m_r g = V_r \rho g, \text{ Barrel: } m_b g > V_b \rho g.$$

Addition of the two relations and comparison with equation shows that:

$$V_r + V_b < V.$$

Therefore, since the volume of the water in the pool is constant, and the total displaced volume is *reduced*, the level of the surface *falls*. This result is perhaps contrary to intuition: since the *whole* volume of the barrel is submerged in (c), it might be thought that the water level will rise above that in (b). However, because the barrel must be heavy in order to sink, the load on the raft and hence V_r are substantially reduced, so that the total displaced volume is also reduced. This problem illustrates the need for a complete analysis rather than jumping to a possibly erroneous conclusion.

PRESSURE CHANGE CAUSED BY ROTATION

Finally, consider the shape of the free surface for the situation shown in Figure (a), in which a cylindrical container, partly filled with liquid, is rotated with an angular velocity *ù*—that is, at N = *ù*/2*ð* revolutions per unit time. The analysis has applications in fuel tanks of spinning rockets, centrifugal filters, and liquid mirrors.

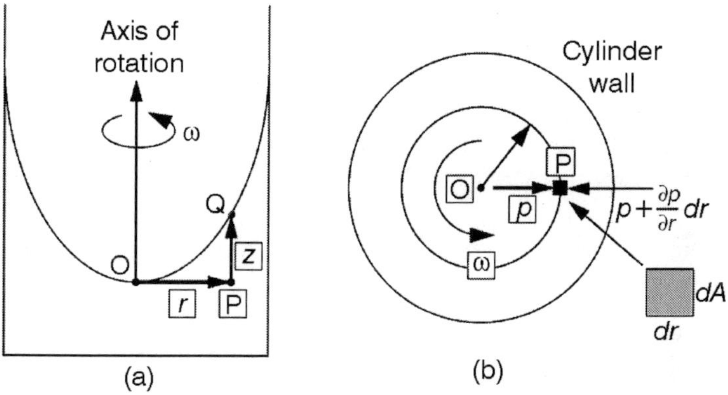

Fig. Pressure changes for rotating cylinder: (a) elevation, (b) plan.

Point O denotes the origin, where $r = 0$ and $z = 0$. After a sufficiently long time, the rotation of the container will be transmitted by viscous action to the liquid, whose rotation is called a *forced vortex.* In fact, the liquid spins as if it were a *solid body*, rotating with a uniform angular velocity *ù*, so that the velocity in the direction of rotation at a radial location *r* is given by $v\theta$ = *rù*. It is therefore appropriate to treat the situation similar to the hydrostatic investigations already made.

Suppose that the liquid element P is essentially a rectangular box with crosssectional area dA and radial extent dr. (In reality, the element has slightly tapering sides, but a more elaborate treatment taking this into account will yield identical results to those derived here.) The pressure on the inner face is p, whereas that on the outer face is $p + (\partial p/\partial r)dr$. Also, for uniform rotation in a circular path of radius r, the acceleration towards the center O of the circle is *rù*2. Newton's second law of motion is then used for equating the net pressure force towards O to the mass of the element times its acceleration:

$$\underbrace{\left(p+\frac{\partial p}{\partial r}dr-p\right)dA}_{\text{Net pressure force}}=\underbrace{\rho(dA\,dr)}_{\text{Mass}}r\omega^2.$$

Note that the use of a *partial* derivative is essential, since the pressure now varies in both the horizontal (radial) *and* vertical directions. Simplification yields the variation of pressure in the radial direction:

$$\frac{\partial p}{\partial r} = \rho r \omega^2,$$

so that pressure *increases* in the radially outward direction. Observe that the gauge pressure at all points on the interface is zero; in particular, $p_O = p_Q = 0$. Integrating from points O to P (at constant z):

$$\int_{p=0}^{p_P} dp = \rho\omega^2 \int_0^r r\,dr,$$

$$p_P = \frac{1}{2}\rho\omega^2 r^2$$

However, the pressure at P can also be obtained by considering the usual hydrostatic increase in traversing the path QP:

$$p_P = \rho g z.$$

Elimination of the intermediate pressure p_P between equations relates the elevation of the free surface to the radial location:

$$z = \frac{\omega^2 r^2}{2g}.$$

Thus, the free surface is *parabolic* in shape; observe also that the density is not a factor, having been canceled from the equations.

There is another type of vortex—the *free* vortex—that is also important, in cyclone dust collectors and tornadoes, for example. There, the velocity in the angular direction is given by $v_¸ = c/r$, where c is a constant, so that $v_¸$ is inversely proportional to the radial position.

OVERFLOW FROM A SPINNING CONTAINER

A cylindrical container of height H and radius a is initially half-filled with a liquid. The cylinder is then spun steadily around its vertical axis Z-Z. At what value of the angular velocity ù will the liquid just start to spill over the top of the container? If $H = 1$ ft and $a = 0.25$ ft, how many rpm (revolutions per minute) would be needed?

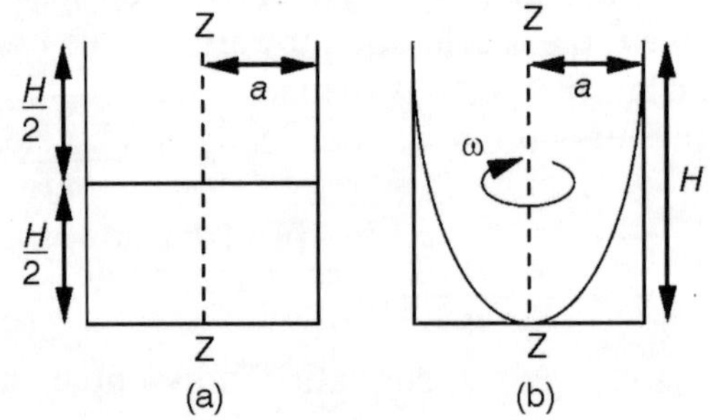

Fig. Geometry of a spinning container:
(a) at rest, (b) on the point of overflowing

Solution: From equation. the shape of the free surface is a parabola. Therefore, the air inside the rotating cylinder forms a paraboloid of revolution, whose volume is known from calculus to be exactly one-half of the volume of the "circumscribing cylinder," namely, the container.8 Hence, the liquid at the center reaches the bottom of the cylinder *just* as the liquid at the curved wall reaches the top of the cylinder. In equation, therefore, set $z = H$ and $r = a$, giving the required angular velocity:

$$\mathrm{w} = \sqrt{\frac{2gH}{a^2}}.$$

For the stated values:

$$\omega = \sqrt{\frac{2 \times 32.2 \times 1}{0.25^2}} = 32.1\frac{\text{rad}}{\text{s}}, \qquad N = \frac{\omega}{2\pi} = \frac{32.1 \times 60}{2\pi} = 306.5 \text{ rpm.}$$

HYDROSTATICS

VARIATION OF PRESSURE WITH ELEVATION.

Here, we investigate how the pressure in a stationary fluid varies with elevation z.

The result is useful because it can answer questions such as "What is the pressure at the summit of Mt. Annapurna?" or "What forces are exerted on the walls of an oil storage tank?" Consider a hypothetical differential cylindrical element of fluid of cross-sectional area A, height dz, and volume Adz, which is also surrounded by the same fluid. Its weight, being the downwards gravitational force on its mass, is $dW = \rho A\, dz\, g$. Two completely equivalent approaches will be presented:

Method 1.

Let p denote the pressure at the base of the cylinder; since p changes at a rate dp/dz with elevation, the pressure is found either from Taylor's expansion or the definition of a derivative to be $p + (dp/dz)dz$ at the top of the cylinder. Hence, the fluid exerts an *upward* force of pA on the base of the cylinder, and a *downward* force of $[p+(dp/dz)dz]A$ on the top of the cylinder. Next, apply Newton's second law of motion by equating the net upward force to the mass times the acceleration—which is zero, since the cylinder is stationary:

$$\underbrace{pA - \left(p + \frac{dp}{dz}dz\right)A}_{\text{Net pressure force}} \underbrace{- \rho A\, dz\, g}_{\text{Weight}} = \underbrace{(\rho A\, dz)}_{\text{mass}} \times 0 = 0.$$

Cancellation of pA and division by Adz leads to the following *differential equation,* which governs the rate of change of pressure with elevation:

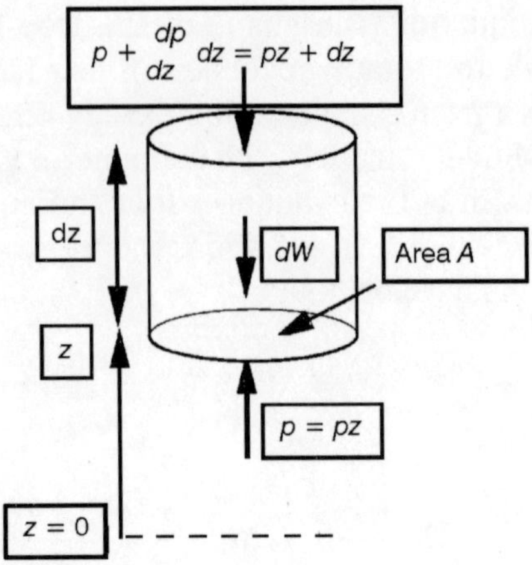

Fig. Forces acting on a cylinder of fluid.

Method 2.

Let pz and $pz+dz$ denote the pressures at the base and top of the cylinder, where the elevations are z and $z+dz$, respectively. Hence, the fluid exerts an *upward* force of pzA on the base of the cylinder, and a *downward* force of $pz+dzA$ on the top of the cylinder. Application of Newton's second law of motion gives:

Isolation of the two pressure terms on the left-hand side and division by Adz gives:

$$\underbrace{p_z A - p_{z+dz} \; A}_{\text{Net pressure force}} - \underbrace{\rho A\,dz\,g}_{\text{Weight}} = \underbrace{(\rho A\,dz)}_{\text{mass}} \times 0 = 0.$$

Isolation of the two pressure terms on the left-hand side and division by Adz gives:

$$\frac{p_{z+dz} - p_z}{dz} = -\rho g.$$

As dz tends to zero, the left-hand side of equation becomes the derivative dp/dz, leading to the same result as previously

$$\frac{dp}{dz} = -\rho g.$$

The same conclusion can also be obtained by considering a cylinder of *finite* height Δz and then letting Δz approach zero.

Note that equation predicts a pressure *decrease* in the vertically upward direction at a rate that is proportional to the local density. Such pressure variations can readily be detected by the ear when traveling quickly in an elevator in a tall building, or when taking off in an airplane. The reader must thoroughly understand *both* the above approaches. For most of this book, we shall use Method 1, because it eliminates the steps of taking the limit of $dz \to 0$ and invoking the definition of the derivative.

Pressure in a liquid with a free surface.

In figure the pressure is ps at the free surface, and we wish to find the pressure p at a depth H below the free surface—of water in a swimming pool, for example.

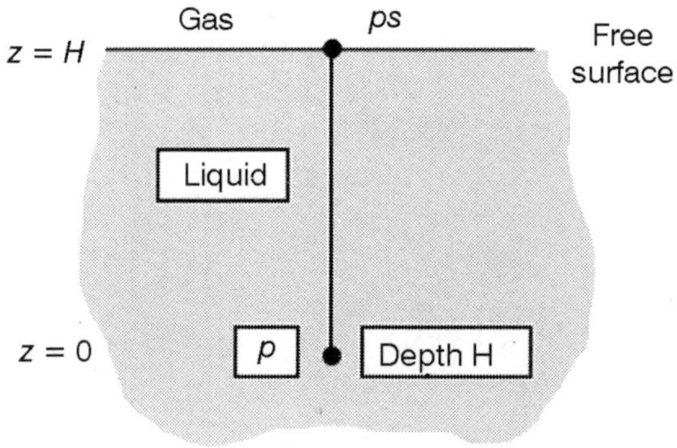

Fig. Pressure at a depth H.

Separation of variables in equation and integration between the free surface ($z = H$) and a depth H ($z = 0$) gives:

$$\int_{ps}^{p} dp = -\int_{H}^{0} \rho g \, dz.$$

Assuming– quite reasonably—that ñ and g are constants in the liquid, these quantities may be taken outside the integral, yielding:

$$p = p_s + \rho g H,$$

which predicts a *linear* increase of pressure with distance downward from the free surface. For large depths, such as those encountered by deep-sea divers, very substantial pressures will result.

EXAMPLE PRESSURE IN AN OIL STORAGE TANK

What is the absolute pressure at the bottom of the cylindrical tank of figure, filled to a depth of H with crude oil, with its free surface exposed to the atmosphere? The specific gravity of the crude oil is 0.846. Give the answers for (a) H = 15.0 ft (pressure in lbf/in2), and (b) H = 5.0 m (pressure in Pa and bar). What is the purpose of the surrounding dike?

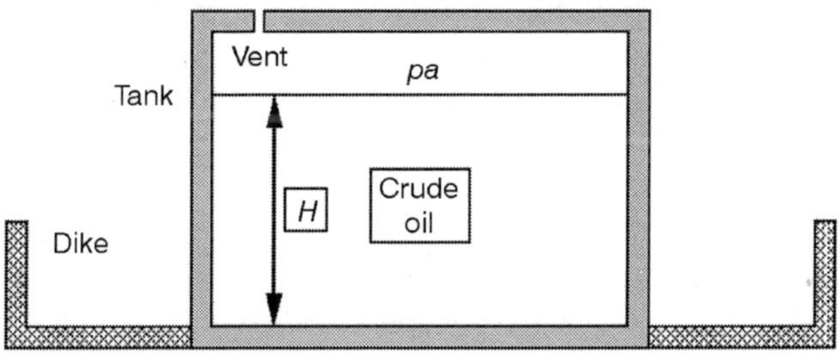

Fig. Crude oil storage tank.

Solution

- The pressure is that of the atmosphere, *pa*, plus the increase due to a column of depth H = 15.0 ft. Thus, setting $p_s = p_a$, equation gives:

$$p = p_a + \rho g H$$
$$= 14.7 + \frac{0.846 \times 62.3 \times 32.215.0}{144 \times 32.2}$$
$$= 14.7 + 5.49 = 20.2 \text{ psia.}$$

The reader should check the units, noting that the 32.2 in the numerator is g [=] ft/s^2, and that the 32.2 in the denominator is gc [=] lbm ft/lbf s^2.

- For SI units, no conversion factors are needed. Noting that the density of water is 1,000 kg/m^3, and that p_a = 1.01 × 10_5 Pa absolute:

$$p = 1.01 \times 10^5 + 0.846 \times 1{,}000 \times 9.81 \times 5.0 = 1.42 \times 10^5 \text{ Pa} = 1.42 \text{ bar.}$$

In the event of a tank rupture, the dike contains the leaking oil and facilitates prevention of spreading fire and contamination of the environment.

Epilogue: When he arrived at work in an oil refinery one morning, the author saw firsthand the consequences of an inadequately vented oil-storage tank.

Rain during the night had caused partial condensation of vapor inside the tank, whose pressure had become sufficiently lowered so that the external atmospheric pressure had crumpled the steel tank just as if it were a flimsy tin can. The refinery manager was not pleased.

EXAMPLE MULTIPLE FLUID HYDROSTATICS

The U-tube shown in figure contains oil and water columns, between which there is a long trapped air bubble. For the indicated heights of the columns, find the specific gravity of the oil.

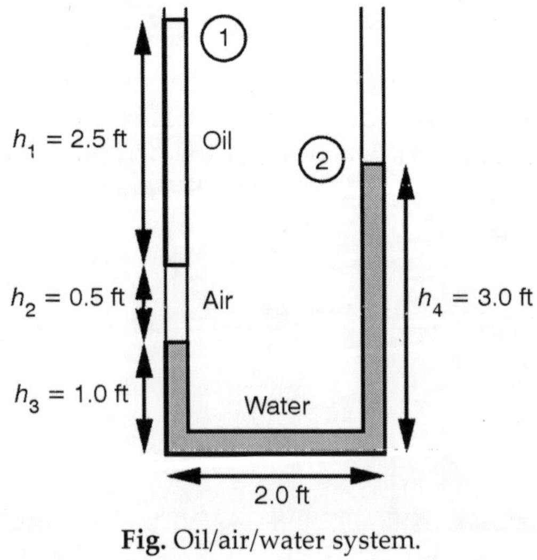

Fig. Oil/air/water system.

Solution: The pressure p^2 at point 2 may be deduced by starting with the pressure p^1 at point 1 and adding or subtracting, as appropriate, the hydrostatic pressure changes due to the various columns of fluid. Note that the width of the U-tube (2.0 ft) is irrelevant, since there is no change in pressure in the horizontal leg. We obtain:

$$p_2 = p_1 + \rho_o g h_1 + \rho_a g h_2 + \rho_w g h_3 - \rho_w g h_4,$$

in which $\tilde{n}_o$, $\tilde{n}_a$, and $\tilde{n}_w$ denote the densities of oil, air, and water, respectively. Since the density of the air is very small compared to that of oil or water, the term containing $\tilde{n}_a$ can be neglected. Also, $p_1 = p_2$, because both are equal to atmospheric pressure. Equation can then be solved for the specific gravity *so* of the oil:

$$s_o = \frac{\rho_o}{\rho_w} = \frac{h^4 - h_3}{h_1} = \frac{3.0 - 1.0}{2.5} = 0.80.$$

Pressure Variations in a Gas.

For a gas, the density is no longer constant, but is a function of pressure (and of temperature—although temperature variations are usually less significant than those of pressure), and there are two approaches:

- For *small* changes in elevation, the assumption of constant density can still be made, and equations similar to equation are still approximately valid.
- For moderate or large changes in elevation, the density in equation is given by, $\rho = M_w p / RT$ or| $\rho = M_w p / Z RT$, depend-ing on whether the gas is ideal or non-ideal. It is understood that *absolute* pressure and temperature must *always* be used whenever the gas law is involved. A separation

of variables can still be made, followed by integration, but the result will now be more complicated because the term *dp/p* occurs, leading—at the simplest (for an isothermal situation)—to a decreasing *exponential* variation of pressure with elevation.

EXAMPLE PRESSURE VARIATIONS IN A GAS

For a gas of molecular weight *M*w (such as the earth's atmosphere), investigate how the pressure p varies with elevation z if $p = p_0$ at $z = 0$. Assume that the temperature T is constant. What approximation may be made for *small* elevation increases? Explain how you would proceed for the non-isothermal case, in which $T = T(z)$ is a known function of elevation.

Solution: Assuming ideal gas behaviour, equation give:

$$\frac{dp}{dz} = -\rho g = -\frac{M_w p}{RT} g.$$

Separation of variables and integration between appropriate limits yields:

$$\int_{p_0}^{p} \frac{dp}{p} = \ln \frac{p}{p_0} = -\int_0^z \frac{M_w g}{RT} dz = -\frac{M_w g}{RT} \int_0^z dz = -\frac{M_w g z}{RT}$$

Since M_{wg}/RT is constant. Hence, there is an exponential decrease of pressure with elevation.

$$p = p_0 \exp\left(-\frac{M_w g}{RT} z\right).$$

Since a Taylor's expansion gives $e^{-x} = 1 - x + x_2/2 - \ldots$, the pressure is *approximated* by $e^{-x} = 1 - x + x^2/2 - \ldots$ the pressure is *approximated* by:

$$p = p_0 \left[1 - \frac{M_w g}{RT} z + \left(\frac{M_w g}{RT}\right)^2 \frac{z^2}{2}\right]$$

For *small* values of $Mwgz/RT$, the last term is an insignificant second-order effect (compressibility effects are unimportant), and we obtain:

$p = p_0 1 - \dfrac{M_w p_0}{RT} gz = p_0 - \rho_0 gz$, in which $\tilde{n}_0$ is the density at elevati-on z = 0; this approximation—essentially one of constant density and is clearly applicable only for a small change of elevation changes the upper limit on z for which this linear approximation is realistic.

If there are significant elevation changes the approximation of equation *cannot* be used with any accuracy. Observe with caution that the Taylor's expansion is only a vehicle for demonstrating what happens for small values of $M_w gz/RT$. Actual calculations for larger values of $M_w gz/RT$ should be made using equation.

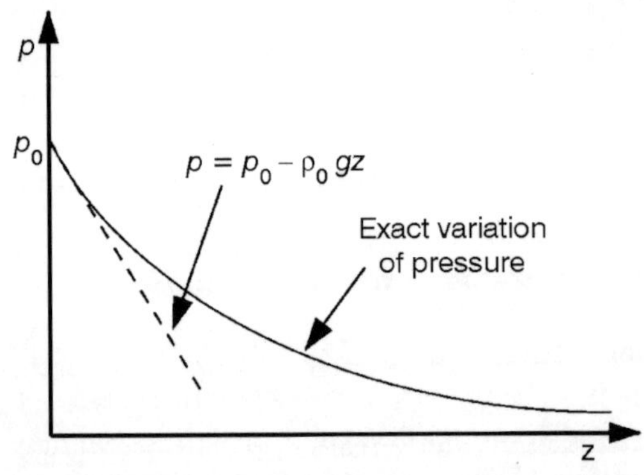

Fig. Variation of gas pressure with elevation.

For the case in which the temperature is not constant, but is a known function $T(z)$ of elevation (as might be deduced from observations made by a meteorological balloon), it must be included *inside* the integral:

$$\int_{p_1}^{p_2} \frac{dp}{p} = -\frac{M_w g}{R} \int_0^z \frac{dz}{T(z)}.$$

Since $T(z)$ is unlikely to be a simple function of z, a numerical method will probably have to be used to approximate the second integral of equation.

Total force on a dam or lock gate.

Figure shows the side and end elevations of a dam or lock gate of depth D and width W. An expression is needed for the total horizontal force F exerted by the liquid on the dam, so that the latter can be made of appropriate strength. Similar results would apply for liquids in storage tanks. Gauge pressures are used for simplicity, with $p = 0$ at the free surface and in the air outside the dam.

Absolute pressures could also be employed, but would merely add a constant atmospheric pressure everywhere, and would eventually be canceled out.

If the coordinate z is measured from the bottom of the liquid upward, the corresponding *depth* of a point below the free surface is $D - z$. Hence, from equation, the differential horizontal force dF on an infinitesimally small rectangular strip of area $dA = Wdz$ is:

$$dF = pWdz = \rho g(D - z)Wdz.$$

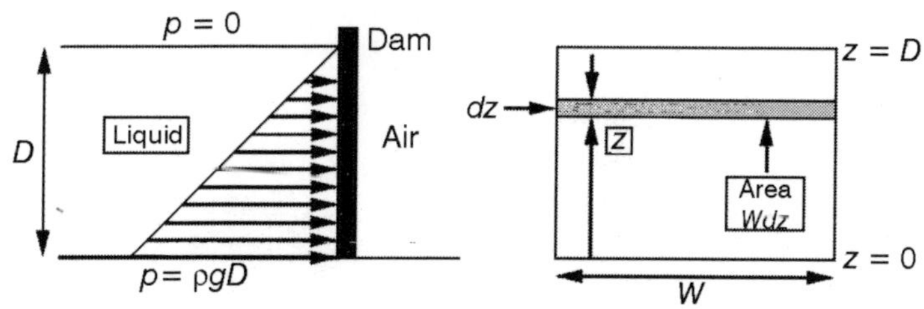

Fig. Horizontal thrust on a dam:
(a) side elevation, (b) end elevation.

Integration from the bottom ($z = 0$) to the top ($z = D$) of the dam gives the total horizontal force:

$$F = \int_0^F dF = \int_0^D \rho g W(D - z)dz = \frac{1}{2}\rho g W D^2.$$

Horizontal pressure force on an arbitrary plane vertical surface.

The preceding analysis was for a regular shape. A more general case, figure shows a *plane* vertical surface of *arbitrary* shape. Note that it is now slightly easier to work in terms of a *downward* coordinate h.

Again taking gauge pressures for simplicity (the gas law is not involved), with $p = 0$ at the free surface, the total horizontal force is:

$$\int_A p dA = \int_A \rho g h dA = \rho g A \frac{\int_A h dA}{A}.$$

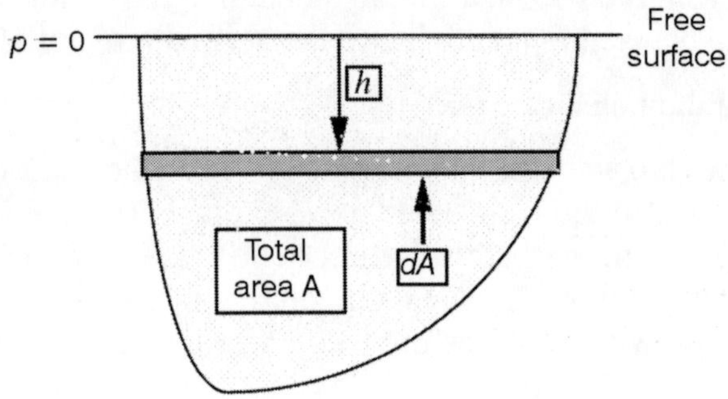

Fig. Side view of a pool of liquid with a submerged vertical surface.

But the depth *hc* of the centroid of the surface is defined as:

$$h_c \equiv \frac{\int_A h dA}{A}.$$

Thus, the total force is: $F = \rho g h_c A = p_c A$, in which *pc* is the pressure at the centroid.

The advantage of this approach is that the location of the centroid is already known for several geometries. For example, for a rectangle of depth *D* and width *W*:

$h_c = \frac{1}{2} D$ and $F = \frac{1}{2} \rho g W D^2$, in agreement with the earlier result of equation. Similarly, for a vertical circle that is just submerged, the depth of the centroid equals its radius. And, for a vertical triangle with one edge coincident with the surface of the liquid, the depth of the centroid equals one-third of its altitude.

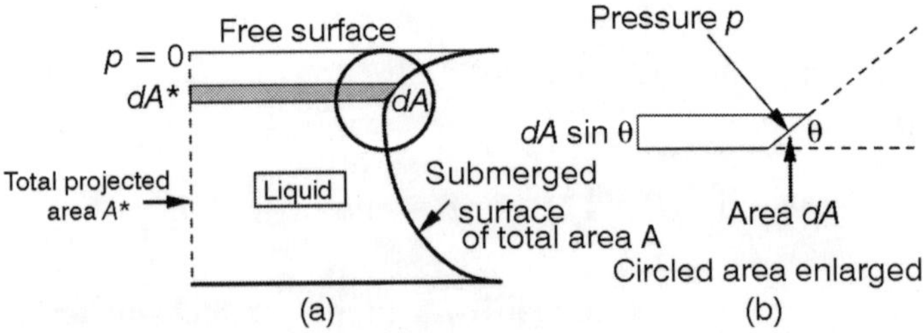

Fig. Thrust on surface of uniform cross-sectional shape.

Horizontal pressure force on a curved surface.

Figure shows the cross section of a submerged surface that is no longer plane. However, the shape *is* uniform normal to the plane of the diagram. In general, the local pressure force *pdA* on an element of surface area *dA* does not act horizontally; therefore, its horizontal component must be obtained by projection through an angle of (ð/2 – è), by multiplying by cos(ð/2 – è) = sin è. The total horizontal force *F* is then:

$$F = \int_A p \sin\theta \, dA = \int_{A^*} p \, dA^*,$$

in which $dA_* = dA$ sin è is an element of the projection of A onto the hypothetical vertical plane A*.

FLUID STATICS

Fluid statics (also called hydrostatics) is the science of fluids at rest, and is a sub-field within fluid mechanics.

The term usually refers to the mathematical treatment of the subject. It embraces the study of the conditions under which fluids are at rest in stable equilibrium. The use of fluid to do work is called hydraulics, and the science of fluids in motion is fluid dynamics.

STATIC PRESSURE IN FLUIDS

Due to an ability to resist deformation, fluids exert pressure normal to any contacting surface. In addition, when the fluid is at rest (static) that pressure is isotropic, *i.e.* it acts with equal magnitude in all directions. This characteristic allows fluids to transmit force through the length of pipes or tubes, *i.e.*, a force applied to a fluid in a pipe is transmitted, via the fluid, to the other end of the pipe. If the force is unequal, the fluid will move in the direction of the resulting force.

This concept was first formulated, in a slightly extended form, by the French mathematician and philosopher Blaise Pascal in 1647 and would be later known as Pascal's law. This law has many important applications in hydraulics. Galileo Galilei, also was a major father of hydrostatics.

Hydrostatic Pressure

Considering a small cube of water at rest below a free surface, the weight of the water above must be balanced by a pressure in this small cube. For an infinitely small cube the stress is the same in all directions and water weight or equivalent pressure can be expressed as,

$$P = \rho g h$$

where, using SI units,

- P is the hydrostatic pressure (in pascals);
- ρ is the water density (in kilograms per cubic metre);
- g is gravitational acceleration (in metres per second squared);
- h is the height of fluid above (in metres).

LIQUIDS-FLUIDS WITH FREE SURFACES

Liquids can have free surfaces at which they interface with gases, or with a vacuum. In general, the lack of the ability to sustain a shear stress entails that free surfaces rapidly adjust towards an equilibrium. However, on small length scales, there is an important balancing force from surface tension.

SURFACE TENSION EFFECTS

Capillary Action

When liquids are constrained in vessels whose dimensions are small, compared to the relevant length scales, surface tension effects become important leading to the formation of a meniscus through capillary action. This capillary action has profound consequences for biological systems as it is part of on of the two driving mechanisms of the flow of water in plant xylem, the transpirational pull.

Drops

Without surface tension, drops would not be able to form. The dimensions and stability of drops are determined by surface tension.

PRESSURE

If a force is exerted on a specific area of an object we define the pressure as the ratio of the force exerted to the area over which it is exerted.

$$P = \frac{F}{A}$$

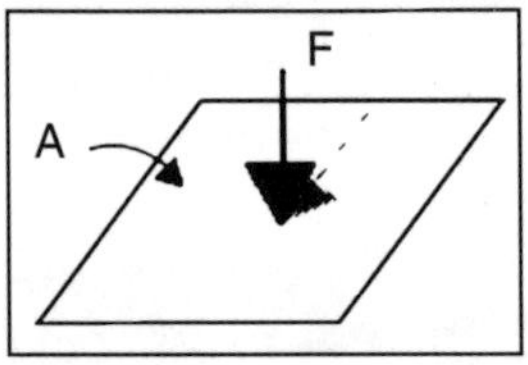

The units of pressure are N/m^2 or the Pascal (Pa). The pressure at sea level on Earth is said to be 1 atmosphere (atm) which equals 1.013×10^5 Pa.

FLUIDS

A fluid is a collection of randomly arranged molecules held together by weak cohesive forces and the forces exerted by the walls of a container. Both gasses and liquids are fluids. Fluids are characterized by their densities.

The pressure in a fluid can be defined as the ratio of the force exerted by the fluid to the area over which it exerted. (To get the pressure at a point you need to take the limit as the area approaches zero.) Because of the weak cohesive forces between the molecules of the fluid the only force that can be applied by a fluid on a object submerged in the fluid is one that tends to compress the object. This means that the force of the fluid is perpendicular to the surface of the object at any point.

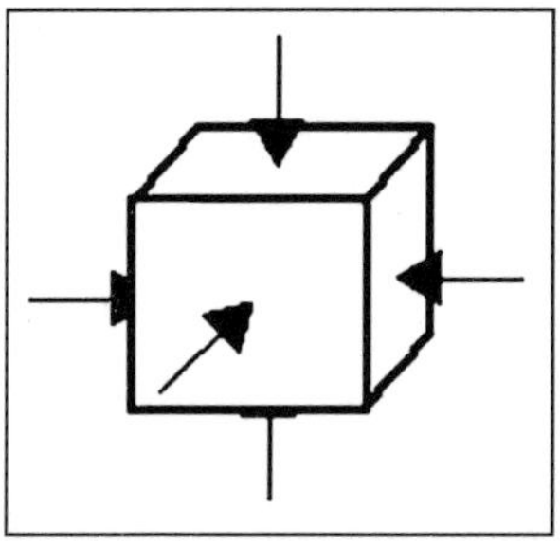

PRESSURE AND DEPTH

It is well known that pressure varies with depth in a fluid. Divers must take precautions against the increased pressure of the ocean depth. Planes flying at high altitudes have pressurized cabins. The relation between absolute pressure and depth near the surface of the Earth can be derived by analyzing the forces on thin area, A, of fluid at some depth, h, in a static fluid. As the fluid is static there is no net force on the thin area of fluid. The forces on the thin area are PA from the pressure below, P_0A from the surface pressure and the weight of the fluid column of depth h.

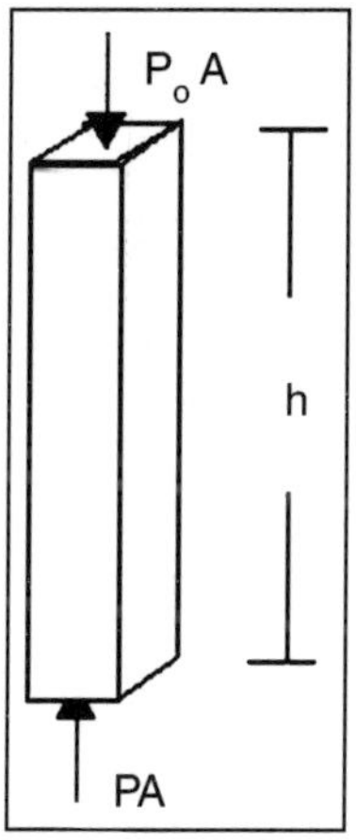

Net Force at bottom of column is,

$PA - P_0A - \rho gAh = 0$

$P = P_0 + \rho gh$

PO is the pressure at the surface of the fluid, (often 1 ATM), h is the depth and P is the pressure at that depth. The quantity P is called the absolute pressure and the difference in pressures, P - PO, is called the gauge pressure.

ARCHIMEDES' PRINCIPLE

We all know that some objects float in water. All objects seem to weigh less when in a fluid. This is due to a buoyant force exerted on the object by the fluid. According to Archimedes' Principle

Any object completely or partially submerged in a fluid experiences a buoyant force, B, equal in magnitude to the weight of the fluid displaced by the object.

This buoyant force acts vertically upward through what was the center of gravity of the displaced fluid.

Let's look at a few cases:

Completely Submerged Object:

- $V_{displaced} = V_{object}$
- $Fnet = B - W = r_{fluid} V_{displaced} g - r_{object} V_{object} g$
- $F_{net} = (r_{fluid} - r_{object}) V_{object} g$

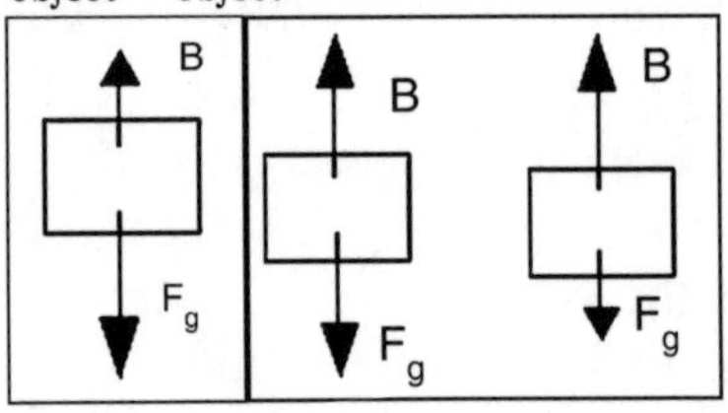

- $\rho_{fluid} < \rho_{object}$
- $\rho_{fluid} = \rho_{object}$
- $\rho_{fluid} > \rho_{object}$

The other possibility is,

Floating Objects:

- $V_{displaced} = V_{submerged}$
- $Fnet = B - W = r_{fluid} V_{submerged} g - r_{object} V_{object} g = 0$
- $r_{fluid} V_{submerged} = r_{object} V_{object}$

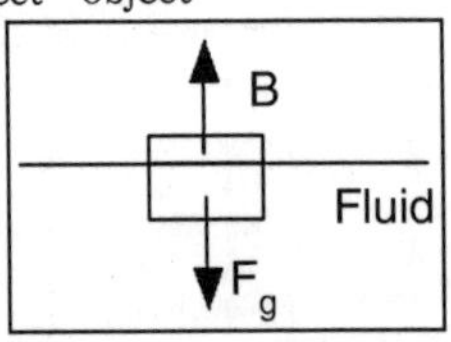

A Worked Example: A giant clam with a mass of 570.0 kg and a volume of 0.43 m^3 lies at the bottom of a freshwater lake. How much force is needed to lift it at constant velocity?

- $F_{net} = 0$
- $F_{net} = F_{lift} + B - mg = 0$
- $F_{lift} = mg - B$
- $B = r_{water} V_{clam} g$
- F_{lift} = (570.0 kg)(9.8 m/s^2) – (1.00 × 103 kg/m^3) (0.43 m^3)(9.8 m/s^2)
- F_{lift} = 1372 N

This is the force needed to lift 140 kg in air.

STREAMLINES AND FLUX

Fluid motion is usually extremely complicated. In order to analyze it we will need to make some assumptions. The ideal fluid flow is,

- *Irrotational*: This basically means the fluid is not turbulent. It flows smoothly.
- *Incompressible*: The density of the fluid is constant.
- *Nonviscous*: The fluid has no internal friction.
- *Steady flow*: The velocity of the fluid at each point is constant in time.

Given these somewhat restrictive conditions we can develop models for how fluids behave. In an ideal fluid flow the path taken by any particle in the fluid is called a streamline. The velocity of the fluid is tangent to the streamline at each point. A set of streamlines is called a flow tube. A simple flow tube is pictured below.

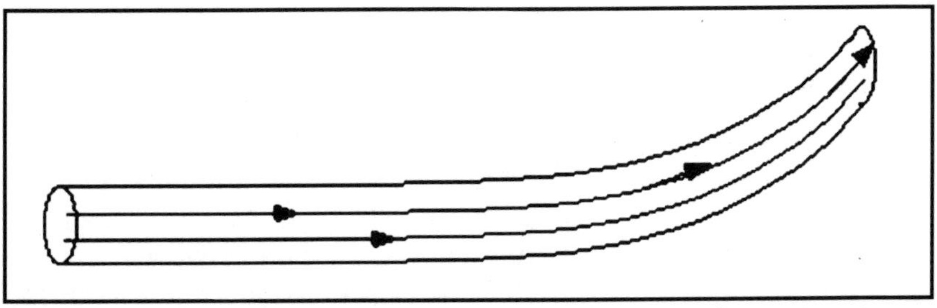

By requiring mass to be conserved in the flow tube it can be shown that the product of the area and the fluid speed at all points along the tube is a constant for an incompressible fluid. The product of the area times the speed is called the flux or flow rate. The relationship is know as the equation of continuity and can be expressed as,

$$A_1 v_1 = A_2 v_2$$

PASCAL'S PRINCIPLE

The fact that pressure varies with depth requires that the pressure at the surface must be transmitted to all parts of the fluid. This was first realized by famous French philosopher and scientist Blaise Pascal.

Pascal's Principle

Any change in pressure applied to a completely enclosed fluid is transmitted undiminished to all parts of the fluid and the enclosing walls.

An example of this is the hydraulic press. (The word "hydraulic" is from the Greek meaning "water tube".) If a force F_1 is applied to a piston A_1 that press on an enclosed fluid Pascal's Principle tells us that a force F_2 will be applied on the much larger area A_2 piston. This allows a small force to be amplified into a much larger force.

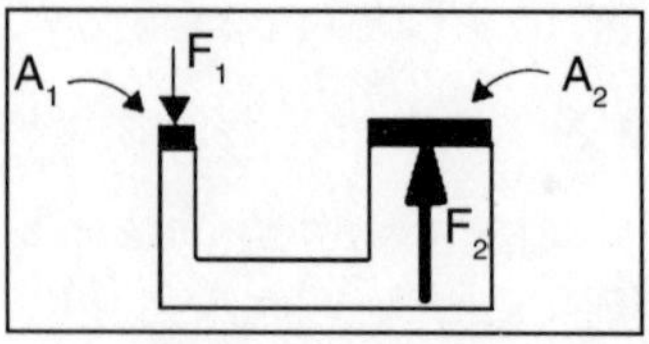

$$F_1/A_1 = F_2/A_2$$

INDUCTION OF ELECTROSTATIC

Electrostatic induction is a redistribution of electrical charge in an object, caused by the influence of nearby charges. Induction was discovered by British scientist John Canton in 1753 and Swedish professor Johan Carl Wilcke in 1762. Electrostatic generators, such as the Wimshurst machine, the Van de Graaff generator and the electrophorus, use this principle. Induction is also responsible for the attraction of light non-conductive objects, such as balloons, paper or styrofoam scraps, to static electric charges. Electrostatic induction should not be confused with electromagnetic induction.

Explanation

A normal uncharged piece of matter has equal numbers of positive and negative electric charges in each part of it, located close together, so no part of it has a net electric charge. The positive charges are the atoms' nuclei which are bound into the structure of matter and are not free to move. The negative charges are the atoms' electrons. In electrically conductive objects such as metals, some of the electrons are able to move freely about in the object.

When a charged object is brought near an uncharged, electrically conducting object, such as a piece of metal, the force of the nearby charge causes a separation of these charges. For example, if a positive charge is brought near the object (see picture at right), the electrons in the metal will be attracted towards it and move to the side of the object facing it. When the electrons move out of an area, they leave an unbalanced positive charge due to the nuclei. This results in a region of negative charge on the object nearest to the external charge, and a region of positive charge on the part away from it. These are called induced charges. If the external charge is negative, the polarity of the charged regions will be reversed.

Since this process is just a redistribution of the charges that were already in the object, it doesn't change the total charge on the object; it still has no net charge. This induction effect is reversible; if the nearby charge is removed, the attraction between the positive and negative internal charges cause them to intermingle again.

Charging an Object by Induction

However, the induction effect can also be used to put a net charge on an object. If, while it is close to the positive charge, the above object is momentarily connected through a conductive path to electrical ground, which is a large reservoir of both positive and negative charges, some of the negative charges in the ground will flow into the object, under the attraction of the nearby positive charge. When the contact with ground is broken, the object is left with a net negative charge.

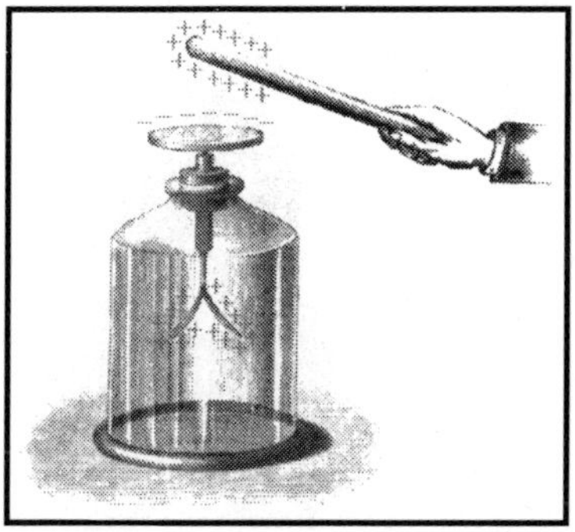

Fig. Gold-Leaf Electroscope, Showing Induction, Before the Terminal is Grounded.

This method can be demonstrated using a gold-leaf electroscope, which is an instrument for detecting electric charge. The electroscope is first discharged, and a charged object is then brought close to the instrument's top terminal. Induction causes a redistribution of the charges inside the electroscope's metal rod, so that the top terminal gains a net charge of opposite polarity to that of the object, while the gold leaves gain a charge of the same polarity. Since both leaves have the same charge, they repel each other and spread apart. The electroscope has not acquired a net charge: the charge within it has merely been redistributed, so if the charge were to be moved away from the electroscope the leaves will come together again.

But if an electrical contact is now briefly made between the electroscope terminal and ground, for example by touching the terminal with a finger, this causes charge to flow from ground to the terminal, attracted by the charge on the object close to the terminal. The electroscope now contains a net charge opposite in polarity to that of the charged object. When the electrical contact to earth is broken, *e.g.* by lifting the finger, the extra charge that has just flowed into the electroscope cannot escape, and the instrument retains a net charge. So the gold leaves remain separated even after the nearby charged object is moved away.

The sign of the charge left on the electroscope after grounding is always opposite in sign to the external inducing charge. On the other hand, an opposite permanent charge on an object can't be achieved if it is grounded from the opposite edge to that which is bearing the external induction charge.

If in the presence of a charged body, an isulated conductor has like charges at one end and unlike charges at the other end this is called the electrostatic induction.

The Electrostatic Field Inside a Conductive Object is Zero

A remaining question is how large the induced charges are. The movement of charge is caused by the force exerted by the electric field of the external charged object. As the charges in the metal object continue to separate, the resulting positive and negative regions create their own electric field, which opposes the field of the external charge. This process continues until very quickly (within a fraction of a second) an equilibrium is reached in which the induced charges are exactly the right size to cancel the external electric field throughout the interior of the metal object. Then the remaining mobile charges (electrons) in the interior of the metal no longer feel a force and the net motion of the charges stops.

Induced Charge Resides on the Surface

Since the mobile charges in the interior of a metal object are free to move in any direction, there can never be a static concentration of charge inside the metal; if there was, it would attract opposite polarity charge to neutralize it. Therefore the mobile charges move under the influence of the external charge until they reach the surface of the metal and collect there, where they are constrained from moving by the boundary.

This establishes the important principle that electrostatic charges on conductive objects reside on the surface of the object. External electric fields induce surface charges on metal objects that exactly cancel the field within. Since the field is the gradient of the electrostatic potential, another way of saying this is that in electrostatics, the potential (voltage) throughout a conductive object is constant.

Induction in Dielectric Objects

A similar induction effect occurs in non-conductive (dielectric) objects, and is responsible for the attraction of small light non-conductive objects, like scraps of paper or Styrofoam, to static electric charges. In non-conductors, the electrons are bound to atoms or molecules and are not free to move about the object as in conductors; however they can move a little within the molecules.

If a positive charge is brought near a non-conductive object, the electrons in each molecule are attracted towards it, and move to the side of the molecule facing the charge, while the positive nuclei are repelled and move slightly to the opposite side of the molecule. Since the negative charges are now closer to the external charge than the positive charges, their attraction is greater than the repulsion of the positive charges, resulting in a small net attraction of the molecule towards the charge. This is called polarization, and the polarized molecules are called dipoles. This effect is microscopic, but since there are so many molecules, it adds up to enough force to move a light object like Styrofoam. This is the principle of operation of a pith-ball electroscope.

Electrostatic Fields

Electrostatic fields are E-fields which do not change with time, which happens when the charges are stationary.

The electric field at a point $E(r)$ is equal to the negative gradient of the electric potential $\Phi(r)$, a scalar field at the same point:

$$E = -\nabla\Phi$$

where ∇ is the gradient. This is equivalent to the force definition above, since electric potential Φ is defined by the electric potential energy U per unit (test) positive charge:

$$\Phi = \frac{U}{q}$$

and force is the negative of potential energy gradient:

$$F = -\nabla U$$

If several spatially distributed charges generate such an electric potential, *e.g.* in a solid, an electric field gradient may also be defined.

Uniform Fields

A uniform field is one in which the electric field is constant at every point. It can be approximated by placing two conducting plates parallel to each other and maintaining a voltage (potential difference) between them; it is only an approximation because of edge effects. Ignoring such effects, the equation for the magnitude of the electric field E is:

$$E = -\frac{\Delta\phi}{d}$$

where Δ? is the potential difference between the plates and d is the distance separating the plates. The negative sign arises as positive charges repel, so a positive charge will experience a force away from the positively charged plate, in the opposite direction to that in which the voltage increases. In micro- and nanoapplications, for instance in relation to semiconductors, a typical magnitude of an electric field is in the order of 1 volt/μm achieved by applying a voltage of the order of 1 volt between conductors spaced 1 μm apart.

Parallels Between Electrostatic and Gravitational Fields

Electric field from a negative Q where

$$F = q\left(\frac{Q}{4\pi\varepsilon_0}\frac{\hat{r}}{|r|^2}\right) = qE$$

Coulomb's law, which describes the interaction of electric charges:

$$F = q\left(\frac{Q}{-4\pi\varepsilon_0}\frac{\hat{r}}{|r|^2}\right) = qE$$

is similar to Newton's law of universal gravitation:

$$F = m\left(-GM\frac{\hat{r}}{|r|^2}\right) = mg .$$

This suggests similarities between the electric field E and the gravitational field g, so sometimes mass is called "gravitational charge".

Similarities between electrostatic and gravitational forces:

- Both act in a vacuum.
- Both are central and conservative.
- Both obey an inverse-square law (both are inversely proportional to square of r).

Differences between electrostatic and gravitational forces:

- Electrostatic forces are much greater than gravitational forces for natural values of charge and mass. For instance, the ratio of the electrostatic force to the gravitational force between two electrons is about 1042.
- Gravitational forces are attractive for like charges, whereas electrostatic forces are repulsive for like charges.
- There are not negative gravitational charges (no negative mass) while there are both positive and negative electric charges. This difference, combined with the previous two, implies that gravitational forces are always attractive, while electrostatic forces may be either attractive or repulsive.

Electrodynamic Fields

Electrodynamic fields are E-fields which do change with time, when charges are in motion.

An electric field can be produced, not only by a static charge, but also by a changing magnetic field. The electric field is given by:

$$E = -\nabla\phi - \frac{\partial A}{\partial t}$$

in which B satisfies

$$B = \nabla \times A$$

and $\nabla\times$ denotes the curl. The vector field B is the magnetic flux density and the vector A is the magnetic vector potential. Taking the curl of the electric field equation we obtain,

$$\nabla \times E = -\frac{\partial B}{\partial t}$$

which is Faraday's law of induction, another one of Maxwell's equations.

Energy in the Electric Field

The electrostatic field stores energy. The energy density u (energy per unit volume) is given by

$$u = \frac{1}{2}\varepsilon |E|^2,$$

where ε is the permittivity of the medium in which the field exists, and E is the electric field vector.

The total energy U stored in the electric field in a given volume V is therefore

$$U = \frac{1}{2}\varepsilon \int_V |E|^2 dV,$$

Further Extensions

Definitive Equation of Vector Fields

In the presence of matter, it is helpful in electromagnetism to extend the notion of the electric field into three vector fields, rather than just one:

$$D = \varepsilon_0 E + P$$

where P is the electric polarization - the volume density of electric dipole moments, and D is the electric displacement field. Since E and P are defined separately, this equation can be used to define D. The physical interpretation of D is not as clear as E (effectively the field applied to the material) or P (induced field due to the dipoles in the material), but still serves as a convenient mathematical simplification, since Maxwell's equations can be simplified in terms of free charges and currents.

Constitutive Relation

The E and D fields are related by the permittivity of the material, ε.

For linear, homogeneous, isotropic materials E and D are proportional and constant throughout the region, there is no position dependence: For inhomogeneous materials, there is a position dependence throughout the material:

$$D(r) = \varepsilon E(r)$$

For anisotropic materials the E and D fields are not parallel, and so E and D are related by the permittivity tensor (a 2nd order tensor field), in component form:

$$D_i = \varepsilon_{ij} E_j$$

For non-linear media, E and D are not proportional. Materials can have varying extents of linearity, homogeneity and isotropy.

HYDROSTATICS FLUIDS

Hydrostatics deals with the laws to which fluids are subjected that do not show motions in hydrostatic coordinate system in which the considerations are carried out, *i.e.* fluids which are at rest in the coordinate system employed for the considerations. General laws of fluid motions they are also applicable to the case of fluids at rest, *i.e.* non-flowing fluids. Thus from the continuity equation

$$\frac{\partial \rho}{\partial t} + \frac{\partial}{\partial x_i}(\rho U_i) = 0$$

it can be shown that for ρ = const and $U_i \neq f(x_i)$ the continuity equation is given by:

$$\underbrace{\frac{\partial \rho}{\partial t} + U_i \frac{\partial \rho}{\partial x_i}}_{D\rho/Dt=0} + \rho \underbrace{\frac{\partial U_i}{\partial x_i}}_{=0} = 0.$$

This means that for $U_i = 0$ the simple partial differential equation holds:

$$\frac{\partial \rho}{\partial t} = 0$$

whose general solution can be stated as follows:

$$\rho = F(x_i).$$

The density ρ in a fluid at rest thus is only a function of the spatial coordinates x_i. When time variations of the density of the fluid occur, these lead inevitably to motions within the fluid because of the relation between the flow and density fields attributable to the continuity equation.

The general equations of momentum can be expressed as:

$$\rho\left[\frac{\partial U_j}{\partial t} + U_i \frac{\partial U_j}{\partial x_i}\right] = -\frac{\partial U_j}{\partial x_i} - \frac{\partial \tau_{ij}}{\partial x_i} + \rho g_j$$

and its special form is deduced for a fluid at rest ($U_j = 0$ and moleculardependent momentum transport $\frac{\partial \tau_{ij}}{\partial x_i} = 0$) to the following system of partial 160

6 Hydrostatics and Aerostatics differential equations. This represents the set of basic equations of hydrostatics and aerostatics:

$$\frac{\partial P}{\partial x_j} = \rho g_j \; (j = 1, 2, 3)$$

or written out for all three directions:

$$\frac{\partial P}{\partial x_1} = \rho g_1, \quad \frac{\partial P}{\partial x_2} = \rho g_2, \quad \frac{\partial P}{\partial x_3} = \rho g_3.$$

In this section the pressure distribution in a fluid, mainly defined by the field of gravity, will be considered more closely. Restrictions are made concerning the possible fluid properties; the fluid is assumed to be incompressible for hydrostatics, i.e ρ = const. This condition is in general quite well fulfilled by liquid, so that the following derivations can be considered as valid for liquids.

For the derivation of the pressure distribution in liquid at rest a rectangular Cartesian coordinate system is introduced, whose position is chosen such that the mass acceleration $\{g_i\}$ given by the field of gravity only shows one component in the negative x_2 direction, *i.e.* the following vector holds

$$\{g_i\} = \{0, -g, 0\}.$$

Then the differential equations given above generally for the pressure can be written as follows:

$$\frac{\partial P}{\partial x_1} = 0, \quad \frac{\partial P}{\partial x_2} = -\rho g, \quad \frac{\partial P}{\partial x_3} = 0.$$

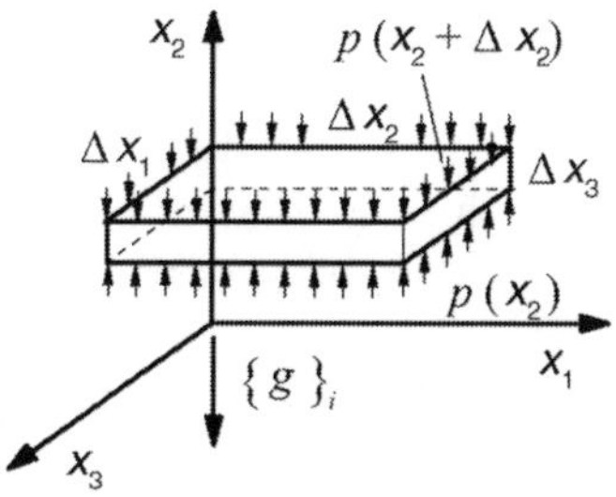

Fig. Coordinate system for the derivation of the pressure distribution in fluids

From $\partial P/\partial x_1 = 0$ follows $P = f\,(x_2\,, x_3)$ and from $\partial P\,/\partial x_3 = 0$ follows $P = f\,(x_1, x_2\,)$. Thus a comparison yields $P = f\,(x_2)$ and this shows that the pressure of a fluid within a plane is constant when the same is vertical to the direction of the field of gravity. The free surface of a fluid stored in a container is a plane of constant pressure and all planes parallel to it are also planes of constant pressure. The pressure increases in the direction that was defined by g_j *i.e.* in the g_j direction of the gravitational acceleration.

For the physical comprehension of hydrostatics it is also important to recognize that equation expresses that the increase in pressure in the negative x_2 direction is caused by the weight of the fluid element plotted in Fig. 6.1 *i.e.*

$$-\rho g \overbrace{\Delta x_1 \Delta x_2 \Delta x_3}^{\Delta V} + p \overbrace{\Delta x_1 \Delta x_3}^{\Delta A} - (p + \frac{\partial p}{\partial x_2} \Delta x_2) \overbrace{\Delta x_1 \Delta x_3}^{\Delta A} = 0.$$

Employing the above physical insights and the resultant equations, the following statements can be made for a liquid of constant density located in a container (Fig.6.2). In the case that the field of gravity acts in the negative x_2 direction, *i.e.*

$$g_1 = 0, g_2 = -g, g_3 = 0$$

the differential equations stated in with the solution $P = f(x_2)$ hold for this case.

Thus the partial differential $\partial P / \partial x_2$ can be written as total differential and one obtains for constant density fluids (ρ =const):

$$\frac{dP}{dx_2} = -\rho g \longrightarrow P - \rho g x_2 + C.$$

or rewritten

$$\frac{P}{\rho} + g x_2 = C$$

This relationship expresses that the sum of the "pressure energy" P / ρ and

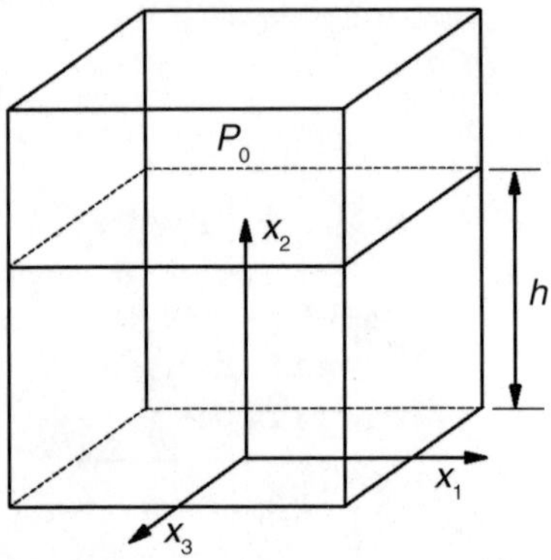

Fig. Fluid at rest in a container

the potential energy ($gx_2 = -g_j x_j$) is a constant at each point of a fluid at rest. As all points of different fluid elements possess the same total energy, the driving force for motion is absent. Thus also from the energy point of view the conditions for hydrostatic fluid behaviour exist.

When the fluid in the height h has a free surface on which an equally large pressure P_0 acts at all points, it represents, because of the relation $P = f(x_2)$, a plane x_2 = const, *i.e.* a horizontal plane.

For the pressure distribution one obtains with the boundary condition $P = P_0$ for $x_2 = h \rightsquigarrow C = P_0 + \rho g h$

$$P = P_0 + \rho g(h - x_2) \; 0 \le x_2 \le h.$$

This relationship expresses the known hydrostatic law, according to which the pressure in a fluid increases in a linear way with the depth below the free surface.

When one rewrites equation one obtains:

$$\frac{P_0}{\rho} + gh = \frac{P}{\rho} + g x_2 = \text{const}$$

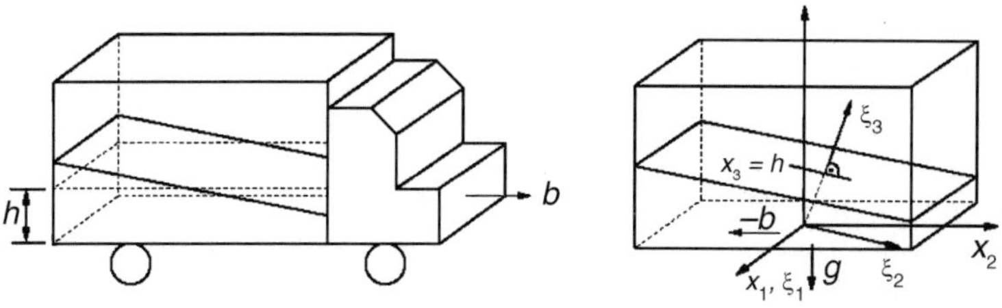

Fig. Position of the fluid level at constant acceleration

The laws of hydrostatics are often applicable also to fluids in moving containers when one treats these as "'accelerated reference systems"'. The externally imposed accelerative forces are then to be introduced as inertia forces. Fig. 6.3 shows as an example, a "container lorry" filled with a fluid which is at rest at the time $t < t_0$ and which increases its speed linearly at for all times $t \geq t_0$, *i.e.* the fluid experiences a constant acceleration.

At a state of rest or in non-accelerating motion, the fluid surface in the container forms a horizontal level. When the container experiences a constant acceleration b, the fluid surface will adopt a new equilibrium position, provided one disregards the initially occurring "'swashing motions"'. When one now wants to compute the new position of the fluid surface, the introduction of a coordinate system x_i, is recommended which is closely connected with the container, where the hydrostatic basic equations read as follows:

$$\frac{\partial P}{\partial x_1} = 0\,; \quad \frac{\partial P}{\partial x_2} = -\rho b\,; \quad \frac{\partial P}{\partial x_3} = -\rho g\,.$$

From this results the general solution:

$$\frac{\partial P}{\partial x_1} = 0 \quad p = f_1(x_2, x_3),$$

$$\frac{\partial P}{\partial x_2} = -\rho b \quad P = -\rho b x_2 + f_2(x_1, x_3),$$

$$\frac{\partial P}{\partial x_3} = -\rho g \quad P = -\rho g x_3 + f_3(x_1, x_2).$$

By comparing the solutions one obtains that f_1, f_2, f_3 can only be the sum of the terms obtained by partial integration plus a constant:

$$P = C - \rho(bx_2 + gx_3).$$

Along the free surface exists the pressure $P = P_0$ and thus the equation of the plane in which the free surface lies reads:

$$x_3 = -\frac{b}{g}x_2 + \frac{1}{g\rho}(C - P_0) \text{ for } -\infty < x_1 < +\infty.$$

The integration constant C is determined by the condition that the fluid volume before and after the onset of the acceleration is the same. Therefore the same relation holds for C as for the container with the fluid at rest:

$$C = g\rho h + P_0.$$

Thus the equation for the plane of the free surface reads:

$$x_3 = h - \frac{b}{g} x_2 \text{ for } -\infty < x_1 < +\infty.$$

As the solution of the problem has to be independent of the chosen coordinate system, a coordinate system ξ_i can be introduced which is rotated against the system x_i in such a way that the following equations for the coordinate transformations hold:

$$\xi_1 = x_1 \text{ (axis of rotation)},$$

$$\xi_2 = \frac{1}{\sqrt{b^2 + g^2}}(gx_2 + bx_3)$$

and

$$\xi_3 = \frac{1}{\sqrt{b^2 + g^2}}(bx_2 + gx_3)$$

This is equivalent to the introduction of a resulting acceleration of the quantity $\sqrt{b^2 + g^2}$ in the direction ξ_3. Thus the basic hydrostatic equations read:

Thus $P = F(x_3)$ holds and $P = \rho\sqrt{b^2 + g^2}\,\xi_3$

The integration constant C results from the boundary condition:

$$P = P_0 + g\rho h \text{ for } \xi_3 = 0$$

$$P = P_0 + \rho g\left(h - \sqrt{1 + \left(\frac{b}{g}\right)^2}\,\xi_3\right)$$

All further statements concerning the problem of the accelerated fluid container can also be made in the coordinate system ξ_i. Along the free surface $P = P_0$ holds and

$$\xi_3 = \frac{h}{\sqrt{1 + \left(\frac{b}{g}\right)^2}}$$

Thus is the equation of the plane in which the free surface lies.

By the above treatment it becomes clear that it is possible to employ the hydrostatic basic laws also in accelerated reference systems, provided the inertia forces are taken into consideration that are attributable to the external motions.

The occurring accelerations (inertial and gravitational) are to be added to a total acceleration in vectorial manner in order to obtain the direction and the quantity of the total acceleration. The free surface appears vertical to the vector of the total acceleration.

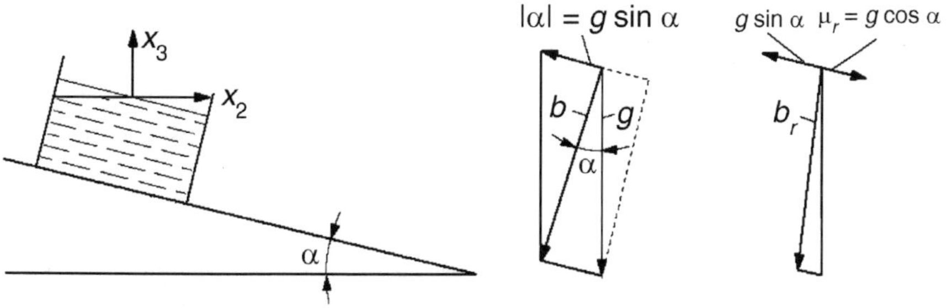

Fig. Water container sliding down on an inclined plane. Motion with and without friction

The example given in Fig. 6.4 can also be categorized into the group of examples that can be treated by means of the basic laws of hydrostatics. This figure shows a water container which is sliding down an inclined plane with an angle of inclination α with respect to the horizontal plane. The container at rest posseses a water surface which is horizontal, as only the gravitational acceleration appears as inertia force per kg of fluid. When the fluid container is released and when the acceleration directed downwards is $| b | = g \sin\alpha$, the body starts moving and experiences in this way an acceleration which is parallel to the inclined plane. The resulting acceleration component acting on the fluid is composed of the component directed upwards with $| b | = g$ sina and the component directed downwards with $\mu_r g \cos \alpha$ Here μ_r is the friction coe cient which characterizes the interaction between the container bottom and the surface of the inclined plane.

When one treats at first the accelerated motion occurring downwards on the inclined plane without friction, one obtains in the coordinate system indicated in Fig. 6.4 the following set of hydrostatic basic equations:

$$\frac{\partial P}{\partial x_1} = 0$$

$$\frac{\partial P}{\partial x_2} = - \rho g \sin \alpha \cos \alpha$$

$$\frac{\partial P}{\partial x_3} = - \rho g(1 - \sin^2\alpha).$$

The position for the pressure distribution in the container sliding downwards in Fig. 6.4 and thus also the solution for the position/location of the fluid surface can be obtained by the solution of the equations.

From $\frac{\partial P}{\partial x_1} = 0$ follows on the one hand $P = f(x^2, x^3)$ and thus the following holds:

$$\frac{\partial P}{\partial x_2} = -\frac{1}{2}\rho g \sin(2\alpha) \longrightarrow P = f_1(x_3) - \frac{1}{2}\rho g \sin(2\alpha) x_2$$

and also

$$\frac{\partial P}{\partial x_3} = -\rho g \cos^2\alpha \longrightarrow P = f_2(x_2) - \rho g(\cos^2\alpha) x_3$$

By comparing the solutions one obtains:

$$P = C - \frac{1}{2}\rho g(\sin(2\alpha)x_2 + 2(\cos^2\alpha)x_3)$$

Along the free surface $P = P_0$ holds and thus one obtains as solution for the location of the free surface:

$$x_3 = (\tan\alpha)x_2 + \frac{1}{\rho g \cos^2\alpha}(C - P_0) \text{ for } -\infty < x_1 < +\infty.$$

As the origin of the coordinates also lies on the free surface $C = P_0$ follows and thus for the plane in which the free surface lies holds:

$$x_3 = -(\tan\alpha)x_2 \text{ for } -\infty < x_1 < +\infty.$$

This equation shows that for friction-free sliding along the inclined plane the free surface lies parallel to the plane along which the container slides. This can also be derived from considerations of the left acceleration diagram in Fig. 6.4 in which it can be seen that the resulting acceleration "'b"' is located vertically with respect to the inclined plane.

When one adds for the downwards motion the occurring frictional force, one obtains the following set of hydrostatic basic equations:

$$\frac{\partial P}{\partial x_1} = 0$$

$$\frac{\partial P}{\partial x_2} = -\rho g(\sin\alpha - \mu_r \cos\alpha)\cos\alpha$$

$$\frac{\partial P}{\partial x_3} = -\rho g[1 - (\sin\alpha - \mu_r \cos\alpha)\sin\alpha].$$

Thus the solution corresponding to the equation reads:

$P = C - \rho g[(\sin\alpha - \mu_r \cos\alpha)\cos\alpha]x_2 - \rho g[1-(\sin\alpha - \mu_r\cos\alpha)\sin\alpha]x_3$.

If one puts on the one hand $P = P_0$, for the free surface, one obtains the equation for the plane in which the free surface lies. When one takes further into consideration, that the origin of the coordinates lies again on the free surface, *i.e.* $C = P_0$, one obtains as final equation:

$$x_3 = -\left[\frac{(\sin\alpha - \mu_r \cos\alpha)\cos\alpha}{1-(\sin\alpha - \mu_r \cos\alpha)\sin\alpha}\right]x_2.$$

For this general case of "friction-loaded motion" along the inclined plane of the fluid container, a free liquid surface appears which is less inclined when compared to the horizontal plane than the inclined plane. Attention has to be paid, however, to the fact that the derivations only hold when $\mu_r \leq \tan\alpha$. For $\mu_r \geq \tan\alpha$ one obtains the limiting case of a container at rest *i.e.* the frictional force is higher then the forward accelerating force.

As a last example to show the employment of hydrostatic laws in accelerated reference systems, the problem presented in Fig. 6.5 is considered. It shows a rotating cylinder closed on the top and at the bottom, which is partly filled with a liquid. When the cylinder is at rest, the free surface of this liquid assumes a horizontal position, as the different liquid particles only experience the gravitational force as mass force. When the cylinder is put into rotation, one observes a deformation of the liquid surface which progresses until as a final form paraboloid.

When now on this rotating motion an additional accelerated vertical motion is superimposed, one detects that the hyperboloid can assume different

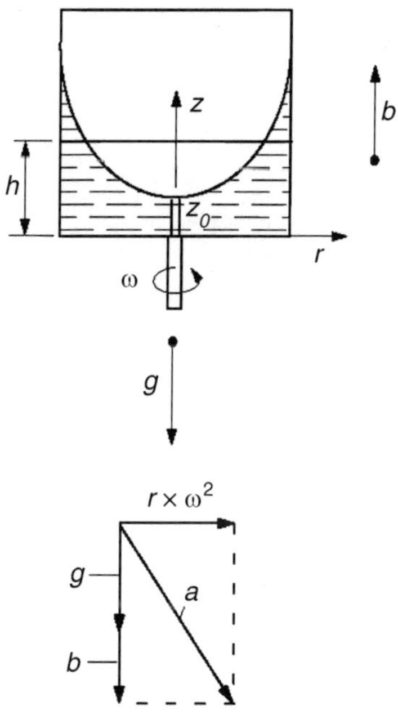

Fig. Treatment of the "'fluid flows"' in a rotating vertically moved and partly filled cylinder

shapes, depending on the magnitude of the vertical acceleration and on the direction in which it takes place. In the following it shall be shown that the issue of the shape of the hyperboloid can be answered on the basis of the basic equations of hydrostatics. For this purpose a coordinate system is chosen, which is firmly coupled to the walls of the rotating and vertically accelerated cylinder and which thus experiences the rotating motion as well as the accelerated vertical motion.

The above mentioned examples have shown that the hydrostatic basic equations are applicable, provided that no fluid motion occurs in the chosen coordinate system and that the external acceleration forces are taken into consideration as inertia forces. In the acceleration diagram sketched in Fig. 6.5 it is shown that for the following derivations the horizontally occurring centrifugal acceleration $\omega^2 r$, as well as the "'vertical acceleration"' b, have been taken into account.

If one considers the processes in the fluid body of the problem sketched in Fig. 6.5, in a coordinate system (r, φ, z), rotating with the cylinder, one finds that all fluid particles are at rest after having reached the stable final state of motion. With reference to the chosen coordinate system the prerequisite for the employment of the hydrostatic basic equation is fulfilled, which in cylindrical coordinates adopt the form indicated below:

$$\frac{\partial P}{\partial r} = \rho g_r \, ; \quad \frac{1}{r}\frac{\partial P}{\partial \varphi} = \rho g_\varphi \, ; \quad \frac{\partial P}{\partial z} = \rho g_z \, .$$

For $g_r = r\omega^2$, $g_\varphi = 0$ and $g_z = -(g + b)$ one obtains for the problem to be treated the following set of basic equations and their general solution.

$$\frac{\partial P}{\partial r} = \rho r \omega^2 \longrightarrow P = \frac{1}{2}\rho\omega^2 r^2 + f_1(\varphi, z)\, ,$$

$$\frac{1}{r}\frac{\partial P}{\partial \varphi} = 0 \longrightarrow P = f_2(r, z),$$

$$\frac{\partial P}{\partial z} = -\rho(g + b) \longrightarrow P = -\rho(g + b)z + f_3(r, \varphi).$$

By comparing the solutions (6.35 a to c) it results in:

$$P = C + \frac{\rho}{2}\omega^2 r^2 - \rho(g + b)z \text{ for } 0 \leq \varphi \leq 2\pi.$$

When one introduces on the axis $r = 0$, for the position of the parabolic apex $z = z_0$, $P = P_0$ holds at the location $r = 0$ and $z = z_0$. This yields for the integration constant:

$$C = P_0 + \rho(g + b)z_0.$$

Therefore the equation for the pressure distribution in the liquid body of Fig. 6.5 reads:

$$P = P_0 + \frac{\rho}{2}\omega^2 r^2 - \rho(g+b)(z-z_0) \text{ for } 0 < \varphi < 2\pi.$$

Along the free surface of the liquid the following holds for the pressure $P = P_0$, so that the free surface employing, can be represented as follows:

$$z = z_0 + \frac{\omega^2}{2(g+b)} r^2 \text{ for } 0 \leq \varphi \leq 2\pi.$$

The introduced apex position z_0 can be determined from the condition that the liquid volume before the rotations starts, i.e $\pi R^2 h$, has to be equal to the liquid volume which exists, in rotation between the free surface of the liquid and the cylinder walls. Thus the following holds:

$$pR^2 h = 2\pi \int_0^R rz\,dr = 2\pi \int_0^R r\left[z_0 + \frac{\omega^2}{2(g+b)} r^2\right] dr$$

and carrying out the integration yields:

$$\frac{1}{2}R^2 h = \left[\frac{1}{2}z_0 r^2 + \frac{\omega^2}{8(g+b)} r^4\right]_0^R = \frac{1}{2}R^2\left[z_0 + \frac{\omega^2}{4(g+b)} R^2\right]$$

$$z_0 = h - \frac{\omega^2}{4(g+b)} R^2 .$$

$$z = h - \frac{\omega^2}{4(g+b)} (R^2 - 2r^2)$$

On the basis of the above indicated relationship the different forms of the free liquid surface can now be looked at. Some typical cases. These will be discussed in the following on the basis of the above derivations and the derived final relationship. It is hoped that it becomes thus clear for the reader how physical information can be obtained by derivations on basic equations of fluid mechanics *e.g.* the form of the free surface of a liquids in containers can be calculated.

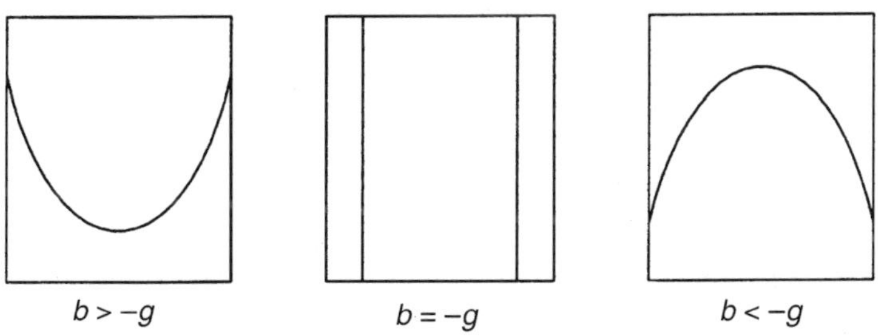

Fig. Examples of possible forms of the fluid surface in a rotating vertically accelerated cylinder

The positions of the liquid surface indicated in Fig. 6.6 can be stated by the indicated relationship in the relative magnitudes of b and g:

$b > -g$: When the vertical acceleration of the container takes place upwards and the resultant b points downwards, respectively, with $0 > b > -g$, the "'opening of the parabola"' is positive according to equation. The liquid touches the bottom and side areas of the container.

$b = -g$: When the vertical acceleration of the container takes place downwards with $b = -g$, the entire fluid rests at the side wall of the container.

$b < -g$: When the vertical acceleration of the container takes place All downwards with $b < -g$, the "'opening of the parabola"' is negative according to equation. The fluid touches the ceiling and side areas of the container.

FREE FLUID SURFACES

SURFACE TENSION

A special characteristic of fluids is that in contrast to solids, they have no form of their own, but always adopt the form of the container in which they are put. While doing this, a free surface forms that the same shows a position which is perpendicular to the vector of the gravitational acceleration.

In this way the fluid properties under gravitational influence were formulated which are known from phenomena of every-day life. It was always assumed that the fluid, at disposal, possesses a total volume having the same order of magnitude as the larger container at disposal. The fluid properties hold only when these conditions are met. This is known from observations of small quantities of liquids which form drops when put on surfaces. It is seen that different shapes of drops can form, depending on

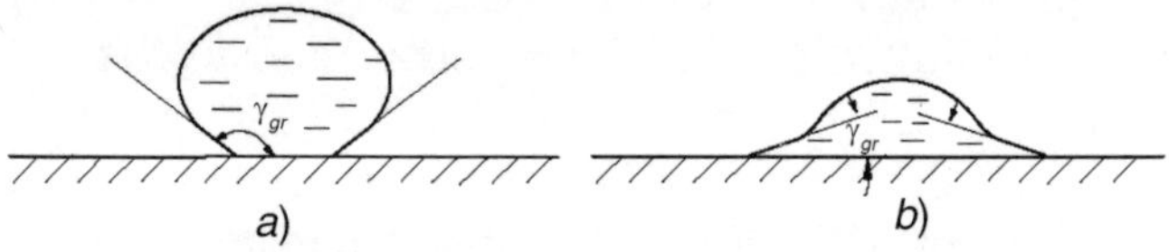

Fig. (a) Shape of Drop in the Case of Non-Wetting Fluid Surfaces; (b) Shape of Drop in the Case of Wetting Fluid Surfaces

which surface and which fluid for forming drops is used. More detailed considerations show moreover that the gas surrounding the fluid and the solid surface all have an influence on the forming shape of a drop. The latter is often neglected and one differentiates considerations of fluid-solid combinations with reference to their wetting possibility, depending on whether the establishing angle of contact between fluid surface and solid surface is smaller than $\pi/2$ or larger.

The surface is classified as non-wetting by the fluid when

$$\gamma_{gr} > \pi/2$$

It holds furthermore that for

$$\gamma_{gr} > \pi/2$$

the surface is classified as wetting for the fluid.

Surfaces covered by a layer of fat are known as examples of surfaces that cannot be wetted by water. Cleaned glass surfaces are to be classified as wetting for many fluids.

The above phenomena can be explained by the fact that different "'actions of forces"' can act on fluid elements. Equivalent physical considerations can be made also owing to the surface energy that can be attributed to free fluid surfaces. When a fluid element is located in a layer that is far away from a free fluid surface, it is surrounded from all sides by homogenous fluid molecules and one can assume that the cohesion forces occurring between the molecules annul each other. This is, however, no longer the case when one considers fluid elements in the proximity of free surfaces. As the forces exerted by gas molecules on the water particles are negligible in comparison to the cohesion forces of the liquid, a particle lying at the free surface experiences an action of forces in direction of the fluid. "'Lateral forces"' also act on the fluid element which thus finds itself in an interphase boundary surface in a state of tension that attributes special characteristics to the free surface. It is thus for example possible to deposit carefully applied flat metal components on free surfaces without fluid penetrating into them. The carrying of razor blades on water surfaces is an experiment that is often presented in basic courses of physics. In nature "'pond skaters"' make use of this particular property of the water surface to cross pools and ponds skillfully and quickly.

When a drop of fluid gets into contact with a firm support adhesion forces also occur in addition to the internal cohesion forces. When these adhesion forces are stronger than the cohesion forces that are typical for the fluid, we have the case of a wetting surface and water drops form. If, however, the cohesion forces are stronger, we have the case of a non-wetting surface and the shapes of the drops.

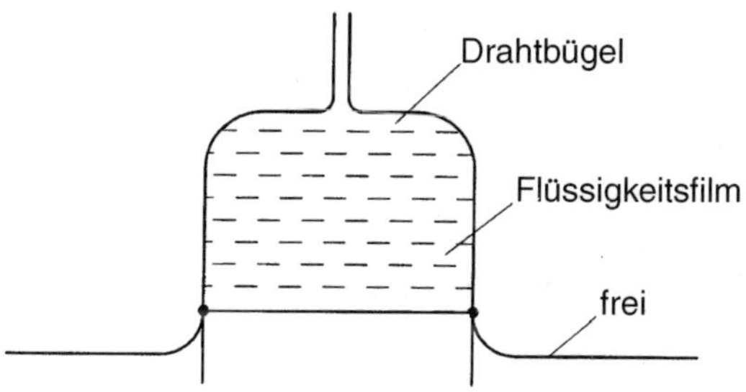

Fig. Strap experiment to prove the action of forces as a consequence of surface tension

More detailed considerations of the processes in the proximity of the free surface of a fluid show that we have to do there with a complicated transit domain (with finite extension vertical to the fluid surface) from a fluid area to a gas area. It suffices, however, for many considerations to be made in fluid mechanics to introduce the surface as a layer with a thickness of $\delta \to 0$. To the same are attributed the properties that comprise the complex transit layers between fluid and gas. The property that is of particular importance for the considerations to be carried out here is the surface tension. This surface tension can be proven by immersing a strap, in a fluid. When pulling the strap through the free surface upwards, one observes that this requires an action of forces which is proportional to the distance between the strap arms. The proportionality constant describing this fact is defined as surface constant.

The surface tension represents thus an action of forces of the free surface per unit linear measure. It can also be introduced as the energy that is required to build up the tension in the liquid film in Fig. 6.14. Both introductions are identical as in both types of energy equation formulated in this way the length of the liquid film in the direction in which the strap is pulled is introduced from the energy set-up. This makes it clear that both possibilities of introduction of the surface tension, one as the action of forces per unit linear measure and other as the energy per unit area, are identical.

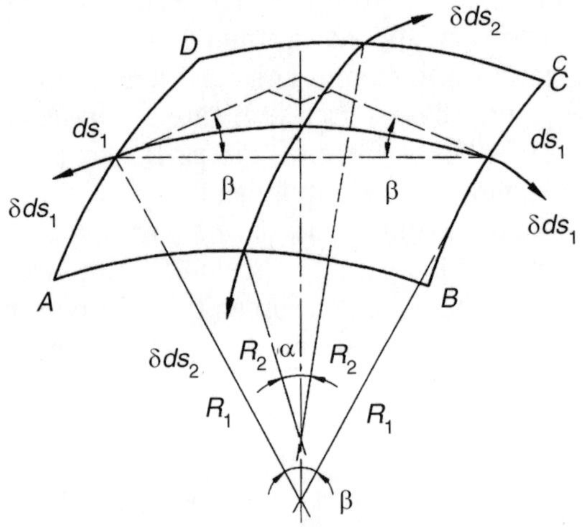

Fig. Schematic representation of a curved surface

In concluding these introductory considerations the e ect of the surface tension on the areas above and below a free surface shall be investigated. From observations of free surfaces in the middle of large containers one can infer that the surface tension there has no influence on the fluid and the gas area lying above it, as the free surface forms vertically to the field of gravity of the earth, as stated in Fig. 6.1. From this it follows that considerations of fluids with free surfaces can be carried out far away from fluid boundaries (container walls) without consideration of the wall effects.

When one considers a curved surface element, one understands that as a consequence of the occurring surface tensions actions of forces are directed to the side of the surface on which the center points of the "'circles of curvature'" are located. The forces attacking on sides AD and BC of the surface element are computed for each element ds_1 and the action of forces resulting from them in direction of the center points of the circles of curvature is:

$$dK_1 = 2\sigma\, ds_1 \sin\beta = 2\sigma ds_1 \frac{ds_2}{2R_1}$$

$$dK_1 = \frac{\sigma}{R_2} ds_1\, ds_2 = \frac{\sigma}{R_1} dO$$

Accordingly the action of forces dK_2 is computed as

$$dK_2 = \frac{\sigma}{R_2} ds_2\, ds_1 = \frac{\sigma}{R_2} dO.$$

This shows that as a consequence of the surface tension pressure effects occur that are directed (in direction of) towards the center points of the circles of curvature. This pressure effect is computed as force per unit area, *i.e.* a differential pressure that is caused by the surface tension:

$$\Delta p_\sigma = \frac{dK_2}{dO} = \sigma\left(\frac{1}{R_1} + \frac{1}{R_2}\right).$$

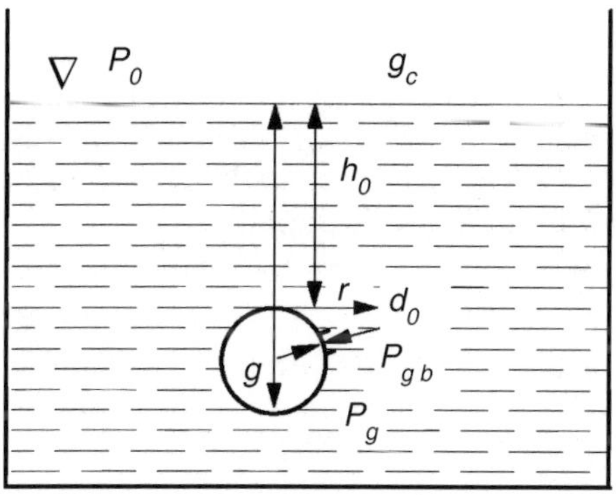

Fig. Diagram for the Consideration of Pressure in Bubbles

When there is a spherical surface it holds

$$R_1 = R_2 = \longrightarrow \Delta p_\sigma = \frac{2\sigma}{R}.$$

This relation means that the gas pressure in a spherical bubble is larger than the fluid pressure imposed from outside:

$$p_F + \frac{2\sigma}{R} = P_g.$$

For very small bubbles this pressure difference can be very large.

When one considers the equilibrium state of a surface element of a bubble, the following relation can be written for the pressure in the upper apex:

$$p_0 + \rho_F g h_0 + \sigma\left(\frac{2}{R_0}\right) = p_{g,0}$$

For a surface element of any height the following pressure equilibrium holds:

$$p_0 + \rho_F g(h_0 + y) + \sigma\left(\frac{1}{R_1} + \frac{1}{R_2}\right) = p_{g,0} + r_g g y.$$

When one now forms the difference of these pressure relations one obtains:

$$\left(\frac{1}{R_1} + \frac{1}{R_2}\right) - \frac{2}{R_2} + \frac{1}{\sigma}(\rho_F + \rho_g) g y = 0\,.$$

Thus the characteristic quantity for the standardization of equation is to be introduced

$$\alpha = \sqrt{\frac{2\sigma}{g(\rho_F - \rho_g)}}\,.$$

which is known as Laplace constant or capillary constant. It has the dimension of a length and indicates in orders of magnitude when a perceptible influence of the surface tension on the surface shape of a medium exists. It holds:

- When the Laplace constant of a free surface of a liquid is comparable with the dimensions of the fluid body, an influence of the surface tension on the fluid shape is to be expected.
- In the proximity of fluid rims (container walls) an influence of the surface tension on the shape of the “fluid surface” is to be expected in areas that are of the order of magnitude of the Laplace constant

Heights of Throw in Tubes and between Plates

From the final statements consequences result for considerations of heights of throw of fluids. Such considerations were carried out, but influences of the boundary surfaces between fluid, solid and gaseous media remained unconsidered there, *i.e.* the influence of the boundary surface tension or surface tension was not taken into consideration. One sees that the considerations stated for communicating systems only hold when the dimensions of the systems are larger than to the Laplace constant of the fluid boundary surfaces. Moreover, the considerations only hold far away from fluid rims. In the immediate proximity of the rim there exists an influence of the surface tension which remained unheeded.

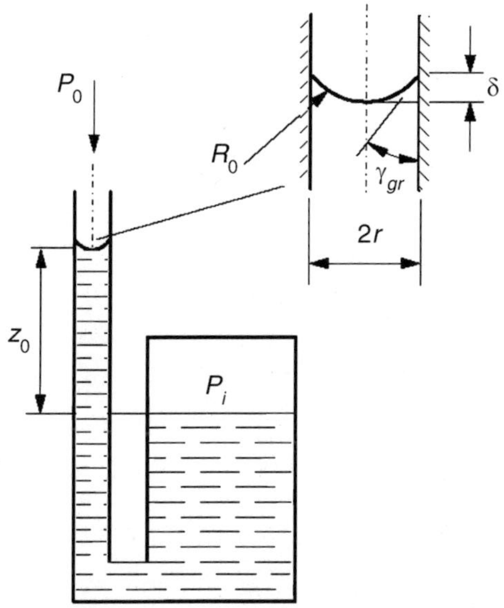

Fig. Diagram for considerations of heights of throw in tubes and between plates

The processes taking place in fluid containers of small dimensions can be treated easily when carrying out a division of the container walls in as "wetting" ones and "non-wetting" ones. When making the considerations at first for wetting walls, experiments show that for such surfaces, in small tubes and between plates with small distances/gaps, the fluid in the tube or between the plates assumes a height which is above the height of the surface of a larger container. This fact is indicated in Fig. 6.17.

From equilibrium considerations it follows:

Pressure between plates $p_0 - \dfrac{\sigma}{R_0} = p_F = p_i - \rho_F g z_0,$

Pressure in tubes $p_0 - \dfrac{2\sigma}{R_0} = p_F = p_i - \rho_F g z_0$

or in other form:

Height of throw between plates $z_0 = \dfrac{1}{\rho_F g}(p_i - p_0) + \dfrac{\sigma}{\rho_F g R_0},$

Height of throw in tube $z_0 = \dfrac{1}{\rho_F g}(p_i - p_0) + \dfrac{2\sigma}{\rho_F g R_0}.$

Here the radius of curvature R_0 is to be considered as an unknown for the determination of which two possibilities exist. To simplify the derivations one can assume with a precision that is sufficient in practice that the surface in the rising pipe adopts the form of a partial sphere for the tube and that of a partial cylinder for the gap of plate. The angle of contact between fluid surface and tube wall or plate wall has to be known from statements on the possibility of wetting. When one defines this angle as γ_{gr}, one obtains the following relation:

$$r = R_0 \cos \gamma_{gr}.$$

For the final relation of the height of throw z_0 for the plates and the tube thus holds:

$$\text{Plates} \qquad z_0 = \frac{1}{\rho_F g}(p_i - p_0) + \frac{\sigma}{\rho_F g r}\cos\gamma_{gr},$$

$$\text{Tube} \qquad z_0 = \frac{1}{\rho_F g}(p_i - p_0) + \frac{2\sigma}{\rho_F g r}\cos\gamma_{gr}.$$

This final relation now shows that even in the case of pressure equality, *i.e.* $p_i = p_0$, the height of throw assumes finite values if $\gamma_{gr} < \frac{\pi}{2}$.

This fact has to be considered when employing communicating systems for measurements of the height of throw and when measuring pressures.

The second possibility to compute the height of pressure is given by the fact that it is experimentally possible, although with a bigger inaccuracy, to determine the quantity δ by means of the following considerations.

$$r^2 + (R_0 - \delta)^2 = R^2_0 \; R_0 = \frac{r^2 + \delta^2}{2\delta}.$$

The height of throw z_0 is computed from this as follows:

$$z_0 = \frac{1}{\rho g}(p_i - p_0) + \frac{4\sigma\delta}{\rho_F g(r^2 + \delta^2)}.$$

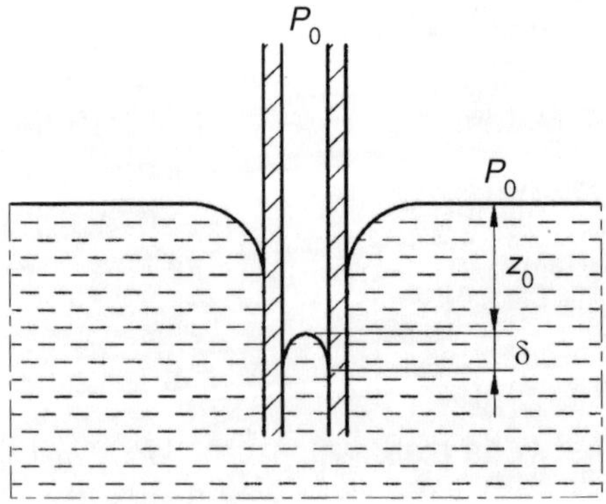

Fig. Considerations of the height of throw in tubes and between plates for non-wetting surfaces

It proves that for $\sigma = 0$ no heights of throw increased by surface e ects are to be expected in tubes or between plates. Under such conditions for the possibility of wetting of the surface the relations hold also for small tube diameters and small gaps between plates.

In the case of non-wetting surfaces it is observed that the fluid in the interior of a rising tube or the gap of a plate does not reach the height which the fluid outside the tube or the gap of the plate assumes. This fact is indicated in Fig. 6.18. Analogous to the preceding considerations for wetting fluids it can be stated:

$$z_0 = \frac{2\sigma}{R_0 \rho g}$$

where R_0 can be introduced again.

The relation thus obtained indicates that the final relations derived for the wetting surfaces can often be applied also to non-wetting media, if one considers the sign of γ_{gr} and d Thus δ is for example to be introduced positively for wetting fluids in the above relations, whereas for non-wetting surfaces d has to be inserted negatively.

Bubble Formation at Nozzles

The injection of gases into fluids for chemical reactions or for an exchangeof-materials represents a process which is employed in many fields of process engineering. Thus bubble formation on nozzles as an introducing process is of interest for these applications. Moreover, the simulation of boiling processes, where the steam bubbles are replaced by gas bubbles, represents another field, where precise knowledge of bubble formation is required.

While gas bubbles form at nozzles during the gassing of liquids, the pressure in the interior of bubbles is changes. For the theoretically conceivable static bubble formation, this is attributed to different curvatures of the bubble boundary surface which are traversed during the formation of bubbles and thus to changes of the capillary pressure. Superimposed upon these are changes in pressure which have their origin in the upward movement of the bubble vertex taking place during the formation. With the dynamic formation of bubbles additional changing pressure e ects are to be expected which are essentially based on accelerative and frictional forces. By static bubble formation one understands the formation of bubbles under pressure conditions, which allow to neglect the pressure effects on an element of the interface boundary surface due to accelerative and frictional forces. Although in practice this is the case only to a very limited extent, the static bubble formation has a certain importance. As it is theoretically conceivable, some important basic knowledge can be gained from it which contributes to the general understanding of bubble formation. Furthermore, knowledge is required on the static bubble formation in order to investigate the influences of the accelerative and frictional forces in the case of dynamic formation of the bubbles.

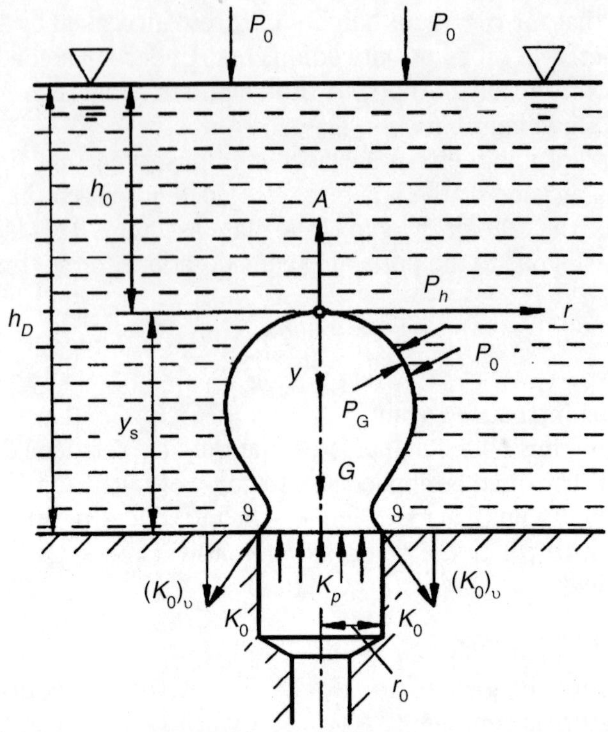

Fig. Equilibrium of forces at a bubble (*A* buoyancy force, *G* gravity, h_D distance of the nozzle from the fluid surface, h_0 distance of the bubble vertex from the fluid surface, K_O surface forces, K_P pressure forces, ph hydrostatic pressure, p_0 atmospheric pressure on the fluid surface)

The essential basic equations of static bubble formation can be derived from the equilibrium conditions for the pressure forces at a boundary-surface ele-ment.

For the pressure equilibrium at an element of the interface boundary surface holds, in accordance with Fig. 6.19, that the gas pressure in the bubble p_G has to be equal to the sum of the hydrostatic pressure p_h and the capillary pressure p_σ

$$p_G = p_\sigma + p_h = \left(\frac{1}{R_1} + \frac{1}{R_2}\right)\sigma + p_0 + \rho_F g(h_0 + y).$$

Here the gas pressure is

$$p_G = p_{G,0} + \rho_G g y$$

When one considers the definition for the radii of curvature, with a as Laplace constant and $\bar{R}_j = R_j / a, \bar{r} = r / a \; \bar{y} = y / a$ the following differ-ential equation can be derived:

$$\frac{\bar{y}''}{(1+\bar{y}'^2)^{3/2}} + \frac{\bar{y}'}{\bar{r}(1+\bar{y}'^2)^{1/2}} = 2\left(\frac{1}{\bar{R}_0} - \bar{y}\right).$$

By the substitution of

$$\overline{z} = \frac{\overline{y}'}{\sqrt{1+\overline{y}'^2}} = \sin\vartheta$$

the differential equation of second order can be replaced by a system of two differential equations of first order

$$\frac{d}{d\overline{r}}(\overline{r}\,\overline{z}) = 2\overline{r}\left(\frac{1}{\overline{R}_0} - \overline{y}\right),$$

$$\frac{d\overline{y}}{d\overline{r}} = \frac{\overline{z}}{\sqrt{1-\overline{z}^2}} = \tan\vartheta,$$

which are used for integration. The desired bubble volume $\overline{V}$ is obtained in dimensionless form by the following partial integration

$$\overline{V} = \pi\int_0^{\overline{y}} \overline{r}^2 d\overline{y} = \pi\overline{r}^2\overline{y} - 2\pi\int_0^{\overline{y}} \overline{r}\,\overline{y}d\overline{r}$$

and with the use of equation

$$\overline{V} = \pi\overline{r}\left[\overline{z} + \overline{r}\left(\overline{y} - \frac{1}{\overline{R}l0}\right)\right].$$

If one introduces again dimension-possessing quantities, the equation can be written as follows:

$$\frac{V}{a^3} = \pi\left(\frac{r}{a}\right)\left[\overline{z} + \frac{r}{a}\left(\frac{y}{a} - \frac{a}{R_0}\right)\right]$$

$$V = a^3\pi r\left[\overline{z} + \frac{r}{a^2}\left(y - \frac{a^2}{R_0}\right)\right].$$

With a and equation the bubble volume V can be written as:

$$V = \frac{2\sigma}{g(\rho_F - \rho_G)}\pi r\left\{\sin\vartheta + \frac{r}{2\sigma}g(\rho_F - \rho_G)\left[y - \frac{2\sigma}{g(\rho_F - \rho_G)R_0}\right]\right\}.$$

Equation represents an integral form of the differential equation system which allows considerations on the equilibrium of forces on bubbles.

For the forces acting on a bubble, the equilibrium condition can be written in the form

$$Vg\rho_F - Vg\rho_G + \pi r^2\left[\frac{2\sigma}{R_0} - g(\rho_F - \rho_G)y\right] = 2\pi r\sigma\sin\vartheta$$

where the first two terms represent the buoyancy force and the weight of the bubble and the third term on the left side is the pressure force on the bubble cross-section πr^2 and the height y The surface forces are indicated on the right side. Equation should be employed in such cases where the bubble volume is to be computed from the conditions of the equilibrium of forces.

For the computation of the pressure changes the pressure in the bubble vertex, owing to a transformation of the equations, can be expressed as stated below:

$$p_{G,0} = \frac{2\sigma}{R_0} + p_0 + \varrho_F g(h_D - y_s);$$

For the pressure at the nozzle mouth varies according to equations

$$p_{G,D} = \frac{2\sigma}{R_0} + p_0 + \varrho_F g h_D - g(\varrho_F - \varrho_g) y_s;$$

Equation can be written in dimensionless form:

$$\Delta \bar{p}_D = \frac{1}{\sqrt{2g\sigma(\varrho_F - \varrho_G)}} [p_{G,D} - p_0 - \varrho_F g h_D] = \frac{1}{R_0} - \bar{y}_s.$$

Although the differential-equation system permits the com-putation of all bubble forms of the static bubble formation and by means of equations, the corresponding bubble volumes and pres-sure differences can be obtained as important quantities of the bubbles, the problem with regard to the single steps of the bubble formation is indefinite/uncertain. The solution of the equations only allows the computation of a one-parameter set of curves, where the vertex radius R_0 is introduced into the derivations as a parameter. It does not permit to predict in which order the different values of the parameter are traversed. This has to be introduced into the considerations as an additional information in order to obtain a set of bubble forms that are traversed in the course of the bubble formation. Theoretically it is now possible to choose any finite, ordered quantity of $R_{0,i}$ values and to compute for these the corresponding bubble forms. Of practical importance, however, is only one $R_{0,i}$ variation, which is given by most of the experimental conditions and for which conditions have been formulated as follows:

- All bubbles form above a nozzle with the radius $\bar{r}_D$.
- $\bar{R}_{0,i}$ = ∞. As starting point of the static bubble formation the horizontal position of the interface boundary surface above the nozzle is chosen.

All further vertex radii are selected according to the condition

$$\bar{V}_D[\bar{R}_{0,i+1}] \geq \bar{V}_D[R_{0,i}]$$

This means that the theoretical investigations are restricted to the bubble formation which comes about through a slow and continuous gas feeding through nozzles having a radius of $\bar{r}_D$. Gas refluxes through the nozzles, and thus a decrease of the bubble volume with mounting vertex radius, as equations would make possible, are excluded by relation c) from the considerations. The consequent application of this relation leads to the formation of a maximum bubble volume. Same has to be considered as volume of the bubble at the start of the separation process, *i.e.*

$$\bar{V}_A = (\bar{V}_D)_{\max}.$$

In the computations the differential equation system was solved numerically for different vertex radii, considering the indicated conditions, and thus the bubble form was ascertained. They can be consulted for the comprehension of the static bubble formation on nozzles in fluids.

Figure shows bubble forms that represent different stages of bubble formation with slow gas feeding through nozzles. The results are reproduced for $\bar{r}_D = 0,4$ and this corresponds to a nozzle radius of $r_D \approx 1,6$ in the case of air bubbles in water.

The change of the bubble volume during the formation of gas bubbles on nozzles of different radii $\bar{r}_D$, where the vertex radius $\bar{R}_0$ was

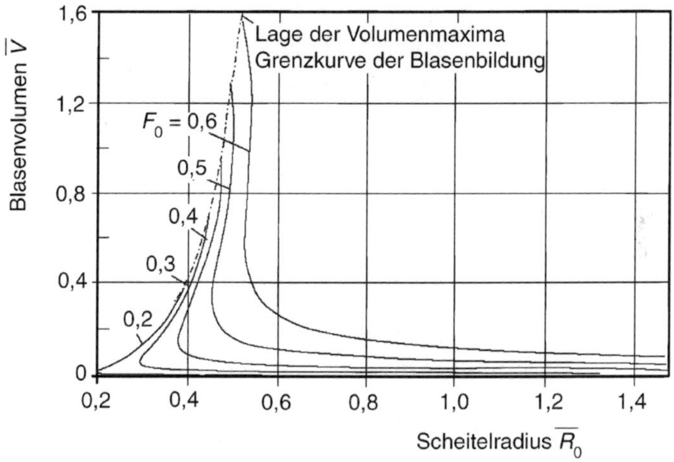

Fig. Bubble forms of the static bubble formation $\bar{r}_D = 0,4$ ascertained through integration of the equation systems

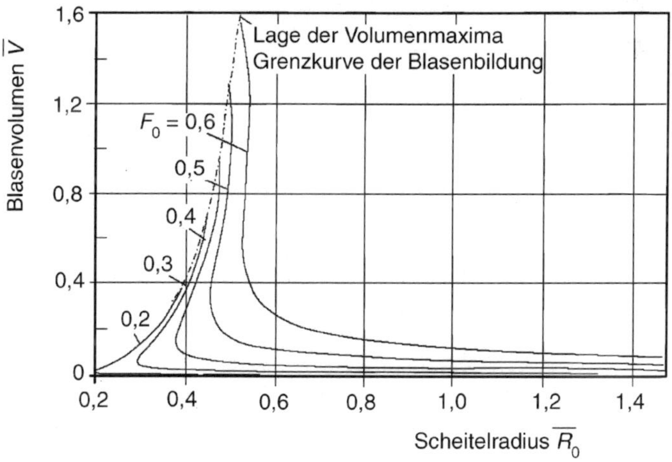

Fig. Bubble volume $\bar{V}$ as a function of the vertex radius $\bar{R}_0$ for the different nozzle radiuses $\bar{r}_D$

chosen for designating the respective formation stage. From this diagram it can be inferred that a large part of the bubble forms at an almost constant vertex radius and it is an important property for larger nozzle radii. For smaller nozzle radii, stronger/larger changes of the vertex radius are to be expected during the formation of the gas bubbles.

The pressure difference $\Delta\overline{P}_D$ as a function of the vertex radius $\overline{R}_0$ is represented for different nozzle radii $\overline{r}_D$ From this representation it can be gathered that for the static bubble formation on nozzles initially a continuous pressure increase at the nozzle mouth is necessary.

After having reached a maximum distinct for all nozzles radii the pressure decreases again. This continuously increasing and then decreasing pressure change, which is required for the static bubble formation, makes it difficult to investigate experimentally the static formation of gas bubbles on nozzles in fluids.

The change of the vertex distance from the nozzle during the bubble formation for different nozzle radii. The vertex radius was chosen for designating the respective stage of bubble formation.

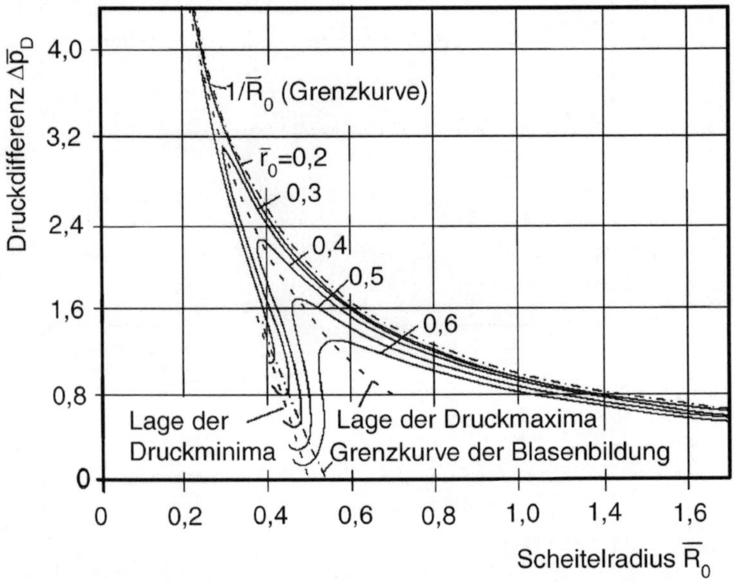

Fig. Pressure difference $\Delta\overline{P}_D$ as a function of the vertex radius $\overline{R}_0$ for different nozzle radiuses $\overline{r}_D$

The dimensionless pressure difference $\Delta\overline{P}_D$ as a function of the bubble volume for different nozzle radiuses $\overline{r}_D$. The final points of the different curves are given by the existence of a maximum bubble volume. The bubble of maximum volume formation above a nozzle is not identical with the one having minimum pressure. Reference is made to the fact, however, that the latter is excluded from the possible bubble forms owing to the definition of the static bubble formation, as it could only be verified by the gas reflux to the nozzles.

For general comprehension of the problem of static bubble formation, two further facts can be referred to:

- For the static bubble formation a basic radius $\overline{r}_{D,gr} = 0.648$, exists which splits up the static bubble formation into two different domains. Theoretical investigations for nozzle radiuses $\overline{r}_D \geq \overline{r}_{D,gr}$ were not carried out

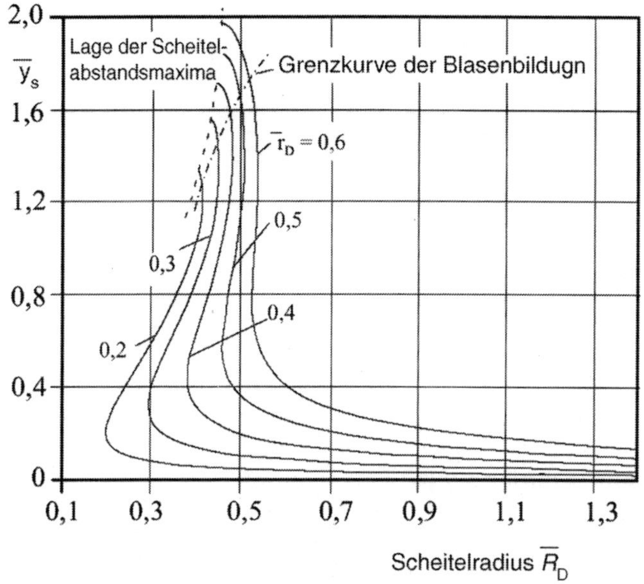

Fig. Distance of the bubble vertex $\overline{y}_s$ ys from the nozzle top as a Function of the vertex radius $\overline{R}_0$ for different nozzle radiuses $\overline{r}_D$

here, as they are not of importance for the introduction of the results intended here for better comprehension of the bubble formation.

- The differential equation system allows the computation of bubble chains. These bubble chains were not investigated further, as they are not in accordance with the above-stated definition of the air feed/ supply for the examined static bubble formation.

While the static nozzle formation can theoretically be understood essentially with simple mathematical means, there exist considerable difficulties for similar investigations of the dynamic bubble formation. This is attribut able mainly to the fact that no coordinate system could be found in which the dynamic bubble formation could be described as a stationary process.

Moreover, for the dynamic bubble formation the pressure on an element of the interface boundary surface is dependent on the fluid motions during bubble formation and thus is computable only by solving the non-stationary Navier-Stokes equation. This, however, is solvable only with difficulty and big computing e orts.

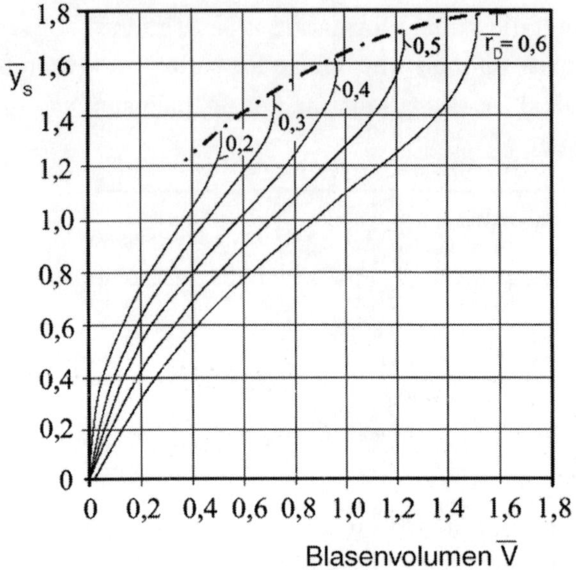

Fig. Distance of the bubble vertex $\overline{y}_s$ from the nozzle top as a function of the nozzle volume $\overline{V}$ for different nozzle radiuses $\overline{r}_D$

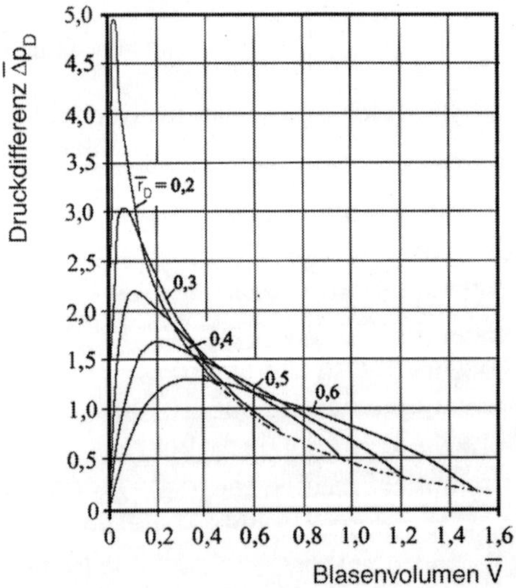

Fig. Pressure difference $\Delta\overline{P}_D$ as a function of the nozzle volume for different nozzle radiuses $\overline{r}_D$

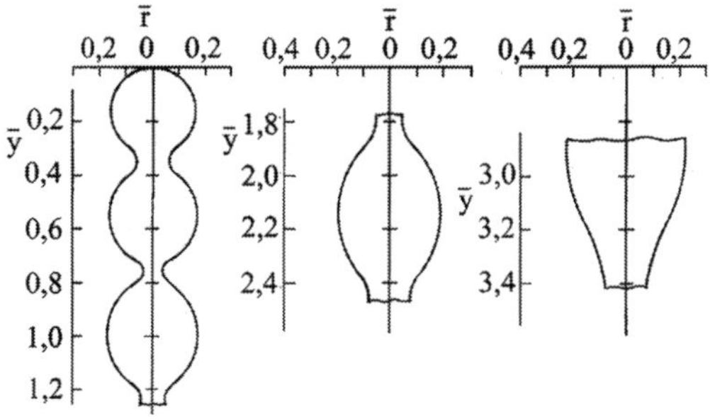

Fig. Bubble chains $\overline{R}_0$ = 1, 60 ascertained from the Differential equation system

5

Hydro-Mechanical Nozzle Flows

The flow problem diagrammed can be solved, neglecting the friction forces, with the aid of the stream-tube theory, in order to obtain in this way a first overview of the flow processes taking place.

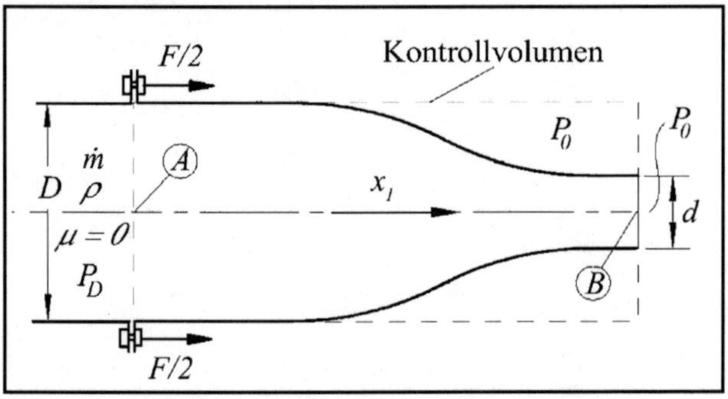

Fig: Nozzle Flow at the End of a Pipe

When applying one-dimensional flow computations, for the flow in cross-section A a velocity can be stated which is constant over the entire tube diameter.

$$U_D = \frac{\dot{m}}{\rho} \frac{1}{\frac{\pi}{4} D^2}$$

Because of the continuity equation holds:

$$\frac{\pi}{4} D^2 U_D = \frac{\pi}{4} d^2 U_d \rightsquigarrow U_d = \frac{D^2}{d^2} U_d$$

From the Bernoulli equation results:

$$\frac{P_D}{\rho} \frac{1}{2} U_D^2 = \frac{P_d}{\rho} + \frac{1}{2} U_2^2 = \frac{P_0}{\rho} + \frac{1}{2} U_2^2$$

From this P_D is computed as:

$$P_D = P_0 + \frac{\rho}{2}\left[U_d^2 - U_D^2\right] = P_0 + \frac{\rho}{2}\left[\left(\frac{D}{d}\right)^4 - 1\right]U_D^2$$

The flange force F on the control volume in Fig. 9.5 is computed from the integral momentum equation as follows:

$$-\rho U_D^2 \frac{\pi}{4} D^2 + \rho U_d^2 \frac{\pi}{4} d^2 - \underbrace{-P_D \frac{\pi}{4} D^2 + P_O \frac{\pi}{4} D^2}_{-\frac{\pi}{4} D^2 (P_D - P_O)} = F$$

or after rearrangement, in consideration of equiation

$$-\frac{\rho}{2}\frac{\pi}{4} D^2 U_D^2 \left[2 - 2\frac{d^2}{D^2}\frac{U_d^2}{U_D^2}\right] - \frac{\pi}{4} D^2 \frac{\rho}{2}\left[\frac{D^2}{d^4} - 1\right]U_D^2 = F$$

and after further rearrangement

$$-\frac{\rho}{2}\frac{\pi}{4} D^2 U_D^2 \left[1 - \frac{D^2}{d^2}\right]^2 = F$$

When inserting the corresponding relations U_D, one obtains for the flange force F :

$$F = -\frac{\dot{m}^2}{\rho \frac{\pi}{2} D^2}\left[1 - \frac{D^2}{d^2}\right]^2$$

DThe force applied by the flange on the examined nozzle part proves to be positive, so that the supporting surface of the flange is pressurized with a negative force F The screws in the flange can therefore be regarded to be force-free.

Sudden Cross-section Widening/Extension

In fluid practice often pipes of different cross-sections are lined up and flows passed/put through them. In this way, viewed in flow direction, internal flows result that are exposed to sudden cross-section widening, as is diagrammed in Fig. 9.6. In this manner separation areas come into being whose influence on the flow shall be made comprehensible by the below-stated considerations.

When carrying out considerations on flows between the planes A and B of the pipe flow, it yields from the continuity equation:

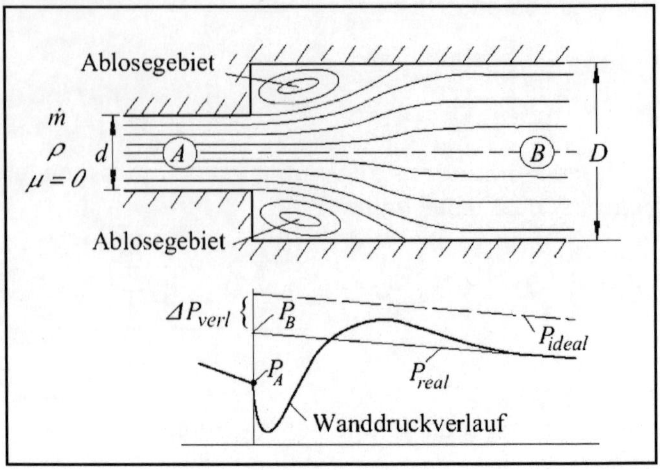

Fig: Carnot's Impact Diffusor

$$\frac{\dot{m}}{\rho} = U_d \frac{\pi}{4} d^2 = U_D \frac{\pi}{4} D^2$$

Thus the velocities U_{d} and U_{D} are known, they are determined by the given mass flow.

In the case of flows passing through the measuring length without losses, the following difference in pressure would result between A and B which can be computed from the Bernoulli equation:

$$\Delta P_{\text{ideal}} = (P_B - P_A)_{\text{ideal}} = \frac{\rho}{2} U_d^2 \left[1 - \frac{d^4}{D^4} \right]$$

Under real conditions, as they are indicated in Fig. 9.6 by the occurrence of the separation areas.

$$F = \rho U_d^2 \frac{\pi}{4} - \rho U_D^2 \frac{\pi}{4} D^2 + P_A \frac{\pi}{4} d^2 - P_B \frac{\pi}{4} D^2$$

When one neglects the contributions to the force F by momentum losses at the pipe walls, then the force F. is computed as the pressure force on the ring surface after the sudden widening, *i.e.* as

$$F = P_A \frac{\pi}{4} (D^2 = d^2)$$

Thus one obtains for the pressure difference

$$DP_{\text{real}} = (P_A - P_B)_{\text{real}} = \rho U_d^2 \frac{d^2}{D^2} \left(1 - \frac{d^2}{D^2} \right),$$

so that a pressure loss (Carnot's impact loss) can be determined as follows:

$$\Delta P_{\text{Verl.}} - \Delta P_{\text{ideal}} = \Delta P_{\text{real.}} = \frac{\rho}{2} U_d^2 \left(1 - \frac{d^4}{D^4}\right) = \frac{\rho}{2}\left(U_d^2 - U_D^2\right)$$

For D $\to \infty$ results as a maximum value for $\Delta P_{\text{Verl}} \cdot = \frac{\rho}{2} U_d^2$, the discharge pressure loss.

INFLUENCE OF AREA CHANGES ON FLOWS

In the present section general considerations shall be carried out which give information on what e ect cross-section changes in flow channels have on fluid flows, *i.e.* to what extent area changes determine the distribution of velocity, pressure, density and temperature along the channel. In the investigation the equations are employed, the continuity equation reads:

$$\tilde{\rho} \tilde{U}_1 F = const$$

Equation can be written in differential form:

$$\frac{dF}{F} + \frac{d\tilde{\rho}}{\tilde{\rho}} + \frac{d\tilde{U}_1}{\tilde{U}_1} = 0$$

The variation of the velocity in flow direction can be described in a first approximation by Euler's equation, reduced for one-dimensional flows, *i.e.* it holds:

$$\tilde{\rho} \tilde{U}_1 \frac{d\tilde{U}_1}{dx_1} - -\frac{d\tilde{P}}{dx_1} = -\frac{d\tilde{P}}{d\tilde{\rho}} \frac{d\tilde{\rho}}{dx_1}$$

On the basis of the energy equation for reversible adiabatic fluid flows, which for area-averaged quantities, by way of approximation, holds as follows:

$$\frac{\tilde{P}}{\tilde{\rho}^x} = const$$

the pressure diversion after the density is to be carried out under adiabatic conditions. As it holds, however:

$$\tilde{c}^2 = \left(\frac{d\tilde{P}}{d\tilde{\rho}}\right)_{\text{ad}}$$

equation can also be written as follows:

$$\tilde{\rho} \tilde{U}_1 \frac{d\tilde{U}_1}{dx_1} = -\tilde{c}^2 \frac{d\tilde{\rho}}{dx_1}$$

When one introduces the Mach number M*a* of the flow as:

$$\widetilde{M_\alpha} = \frac{\tilde{U}_1}{\tilde{c}}$$

equation can be written:

$$\frac{d\tilde{\rho}}{\tilde{\rho}} = -\widetilde{M_\alpha}^2 \frac{d\tilde{U}_1}{\tilde{U}_1}$$

Inserting the result, we obtain:

$$\frac{dF}{F} - \widetilde{M_\alpha}^2 \frac{d\tilde{U}_1}{\tilde{U}_1} + \frac{d\tilde{U}_1}{\tilde{U}_1} = 0$$

or

$$\frac{d\tilde{U}_1}{\tilde{U}_1} = \frac{1-}{\left(1-\widetilde{M_\alpha}^2\right)} \frac{dF}{F}$$

When one takes into consideration that subsonic flows are given by $\widetilde{M_\alpha} < 1$ and supersonic flows by $\widetilde{M_\alpha} > 1$ the above relation expresses:

- In the presence of a subsonic flow ($\widetilde{M_\alpha} < 1$). a decrease of the cross-sectional area of a flow channel in flow direction is linked to an increase of the flow velocity. An increase of the channel cross-sectional area in flow direction results in a decrease of the flow velocity.
- In the presence of a supersonic flow ($\widetilde{M_\alpha} > 1$) a decrease of the cross-sectional area of a flow channel in flow direction is liked to a decrease of the flow velocity. An increaseincrease of the flow cross-section in flow direction results in an increase of the flow velocity.

Besides the changes of the flow velocity caused by changes of the cross-sectional areas, the changes in pressure, density and temperature of the flowing fluid are also of interest. From equation can be seen that the relative change in density has/owns the opposite sign of the change in velocity, *i.e.* the density increases in flow direction when the velocity decreases/drops o and inversely. In the area of subsonic flow the locally present relative change in density is smaller than the local relative change in velocity. In the area of supersonic flow the locally present relative change in density is larger than the relative change in velocity. As concerns the dependence from the cross- sectional area changes of the flow channel, it results for the change in density:

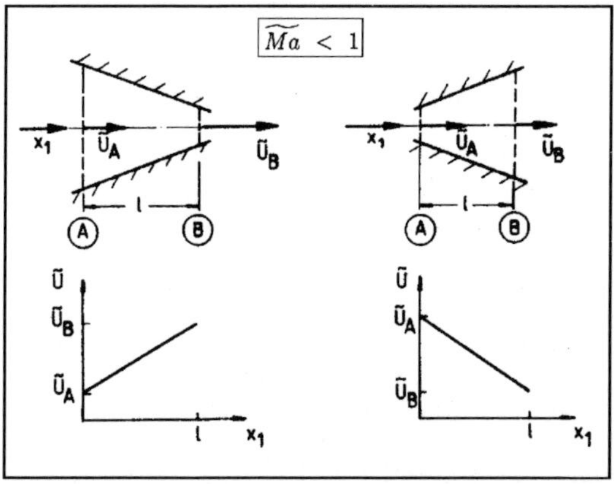

Fig: Influence of the Change of the Flow Cross-Section on a Subsonic Flow

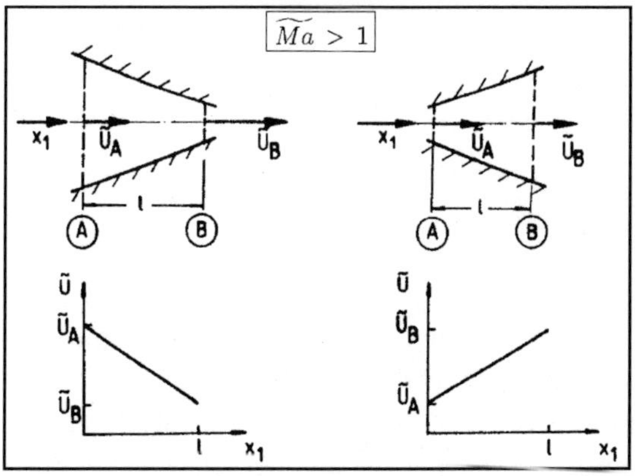

Fig: Influence of the change of the Flow Cross-Section on a Supersonic flow

$$\frac{d\tilde{\rho}}{\tilde{\rho}} = \frac{\widetilde{M}_{\alpha}^{2}}{\left(1-\widetilde{M}_{\alpha}^{2}\right)} \frac{dF}{F}$$

With regard to the pressure variation the following considerations can be carried out. From the adiabatic equation follows:

$$d\tilde{P} = \frac{\tilde{P}}{\tilde{\rho}^{\kappa}} \kappa_{\tilde{\rho}(\kappa-1)} d\tilde{\rho} = \kappa \frac{\tilde{P}}{\tilde{\rho}} d\tilde{\rho}$$

Thus it holds for the local relative change in pressure

$$\frac{d\tilde{P}}{\tilde{P}} = \kappa \widetilde{M}_{\alpha}^{2} \frac{d\tilde{U}_1}{\tilde{U}_1}$$

or with regard to the local relative change of the cross-sectional area of the flow:

$$\frac{d\tilde{P}}{\tilde{P}} = \frac{\kappa \widetilde{M_\alpha}^2}{\left(1-\widetilde{M_\alpha}^2\right)} \frac{dF}{F}$$

Finally it is necessary to consider the variations in temperature. To this end the state equation for ideal gases is differentiated:

$$-\tilde{P}\frac{d\tilde{\rho}}{\tilde{\rho}^2} + \frac{d\tilde{P}}{\tilde{\rho}} = Rd\tilde{T}\frac{dF}{F}$$

or transcribed:

$$-\frac{d\tilde{\rho}}{\tilde{\rho}} + \frac{d\tilde{P}}{\tilde{P}} = \frac{d\tilde{T}}{\tilde{T}}$$

Thus follows from the preceding relations

$$\frac{d\tilde{T}}{\tilde{T}} = -(\kappa - 1)\widetilde{M_\alpha}^2 \frac{d\tilde{U}_1}{\tilde{U}_1}$$

The locally occurring relative change in temperature has the opposite sign of the local relative change in velocity. The occurring relative changes in temperature are weaker than the corresponding relative changes in density. With regard to the relative area change of the flow cross-section it results:

$$\frac{d\tilde{T}}{\tilde{T}} = \frac{(\kappa - 1)\widetilde{M_\alpha}^2}{\left(1-\widetilde{M_\alpha}^2\right)} \frac{dF_1}{F}$$

The considerations stated for the flow-velocity variation in supersonic and subsonic flows, which are sketched in Figs. 9.7 and 9.8 can also be carried out for the variations in pressure, density and temperature with the aid of the above equations.

Another important consideration can be stated through rearrangement of the above-derived relations such that it holds:

$$\frac{dF}{d\tilde{U}_1} = \frac{F}{\tilde{U}_1}\left(1-\widetilde{M_\alpha}^2\right)$$

This relation expresses that the condition for achieving the sound velocity is given by $dF = 0$, *i.e.* $\widetilde{M_\alpha} = 1$ As for the second derivation holds:

$$\frac{\mathrm{d}^2 F}{\mathrm{d}\tilde{U}_1^2} = \frac{F}{\tilde{U}_1^2} \widetilde{M_\alpha}^2 \left(\widetilde{M_\alpha}^2 - 2 \right)$$

for $\widetilde{M_\alpha} = 1$ holds a minimum of the flow cross-section.

PRESSURE-DRIVEN COMPENSATING FLOWS THROUGH CONVERGING NOZZLES

In many technical plants flows of gases occur which are to be classified into the large group of compensating flows that can take place between reservoirs with differing pressure levels. Thus gases *e.g.* are often stored under high pressure in large storage reservoirs, in order to be led via correspondingly dimensioned/designed openings with connecting aggregates and discharge conduits to the intended purpose when need arises. This discharge can idealized be understood as a compensating flow between two reservoirs or two chambers of which one represents the storage reservoir under pressure, while the environment represents the second reservoir. In the following considerations it is assumed that both reservoirs are very large so that constant reservoir conditions exist during the entire compensating flow under investigation. These are assumed to be known and are given by the pressure P_H, the temperature T_H etc. in the high-pressure reservoir, as well as through the pressure

P_N or. T_N for the low-pressure reservoir. The compensating flow shall take place via a continually converging nozzle, whose largest cross-section represents thus the discharge opening of the large reservoir, whereas the smallest nozzle cross-section represents the entrance/inlet opening into the low-pressure reservoir.

When one wants to investigate the fluid flows taking place in the above compensating flow more in detail, the final equations for flows through channels, pipes etc.

$$\tilde{\rho}\tilde{U}_1 F = const$$

$$\tilde{h} + \frac{1}{2}\tilde{U}_1^2 = \text{const}; \quad \frac{\tilde{P}}{\tilde{\rho}^\kappa} = \text{const}$$

$$\frac{\tilde{P}}{\tilde{\rho}} = \mathrm{R}\tilde{\mathrm{T}}$$

With that a su cient number of equations exists to determine the course of the area-averaged velocity and the area-averaged thermo-dynamical state quantities of the flowing gas during the process of the compensating flow, *i.e.* along the x_1 —axis stated in Fig.

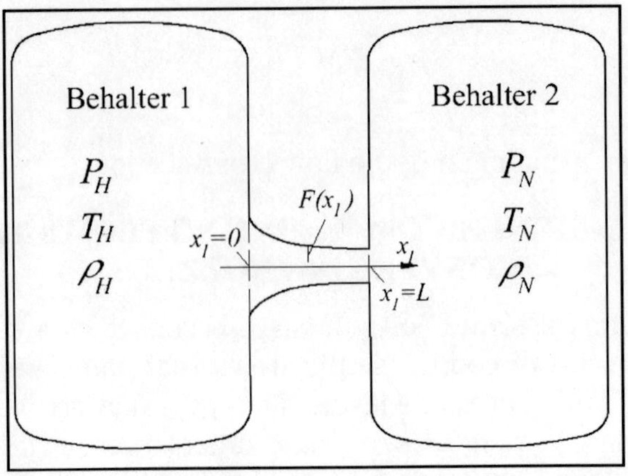

Fig: Compensating Flow Between two Reservoirs through Converging Nozzle

When one considers that - based on the assumption of a large reservoir - in the interior of the high-pressure reservoir there is the constant pressure P_H and the velocity $(U_1)_H = 0$ then for the velocity U_1 at each point x_1 of the nozzle the following relation can be stated:

$$\tilde{h} + \frac{1}{2}\tilde{U}_1^2 = h_H$$

Taking into account that the enthalpy for an ideal gas can be stated as c_P T and that moreover the ideal gas equation holds, the above relation can be transcribed as follows:

$$c_P \frac{\tilde{P}}{R\tilde{\rho}} + \frac{1}{2}\tilde{U}_1^2 = \frac{\kappa}{\kappa - 1}\frac{\tilde{P}}{\tilde{\rho}} + \frac{1}{2}\tilde{U}_1^2 = \frac{\kappa}{\kappa - 1}\frac{P_H}{\rho_H}$$

The velocity U_1. is thus linked to the course of the pressure along the axis of the nozzle as follows:

$$\tilde{U}_1 = \sqrt{\frac{2\kappa}{\kappa - 1}\left(\frac{P_H}{\rho_H} - \frac{\tilde{P}}{\tilde{\rho}}\right)}$$

The above equation indicates that for $\tilde{p} = 0$, *i.e.* for the outflow into a˜ vacuum, a maximum flow velocity develops which is given by the state of the reservoir only:

$$U_{\max} = \sqrt{\frac{2\kappa}{\kappa - 1}\frac{P_H}{\rho_H}} = \sqrt{2c_P \cdot T_H}$$

Standardizing the flow velocity U_1, with $U_{\max}$, existing at a Point x_1, one obtains:

$$\frac{\tilde{U}_1}{U_{max}} = \sqrt{1 - \frac{\tilde{P} \cdot \rho_H}{P_H \cdot \tilde{\rho}}}$$

or transcribed by means of the ideal gas equation:

$$\frac{\tilde{U}_1}{U_{max}} = \sqrt{1 - \frac{\tilde{T}}{T_H}}$$

Linking the adiabatic equation to the state equation the following relations:

$$\frac{\tilde{T}}{T_H} = \left(\frac{\tilde{\rho}}{\rho_H}\right)^{\kappa-1} \quad \textit{and} \quad \frac{\tilde{T}}{T_H} = \left(\frac{\tilde{P}}{P_H}\right)^{\frac{\kappa-1}{\kappa}}$$

Thus the following equations hold:

$$\frac{\tilde{U}_1}{U_{max}} = \sqrt{\left[1 - \left(\frac{\tilde{\rho}}{\rho_H}\right)^{\kappa-1}\right]}$$

and

$$\frac{\tilde{U}_1}{U_{max}} = \sqrt{\left[1 - \left(\frac{\tilde{P}}{P_H}\right)^{\frac{\kappa-1}{\kappa}}\right]}$$

When choosing the standardized velocity $(\tilde{U}_1 / U_{max})$ as a parameter for the representation of the flow in the nozzle, the course of pressure, density and temperature can be stated as follows:

$$\frac{\tilde{P}}{P_H} = \left[1 - \left(\frac{\tilde{U}_1}{U_{max}}\right)^2\right]^{\frac{\kappa}{\kappa-1}}$$

$$\frac{\tilde{\rho}}{P_H} = \left[1 - \left(\frac{\tilde{U}_1}{U_{max}}\right)^2\right]^{\frac{\kappa}{\kappa-1}}$$

$$\frac{\tilde{T}}{T_H} = \left[1 - \left(\frac{\tilde{U}_1}{U_{max}}\right)^2\right]$$

These relations are sated in Fig. 9.10 as functions of $(\tilde{U}_1 / U_{max})$ Also stated is, along the $(\tilde{U}_1 / U_{max})$–axis, the corresponding Mach number of the flow, which in consideration of the relation $c = c = \sqrt{(\mathrm{d}P/\mathrm{d}\rho)_{\text{ad}}} = \sqrt{\kappa RT}$ can be computed as follows:

$$\frac{\tilde{U}_1^2}{U_{\max}^2} = \frac{\tilde{U}_1^2}{2c_P T_H}\frac{\kappa R\tilde{T}}{\kappa R\tilde{T}} = \widetilde{M_{\alpha 1}}^2 \frac{\kappa-1}{2}\left(\frac{\tilde{T}}{T_H}\right)$$

When one considers the relation derived above for (T/T_H) equation, one obtains for the Mach number to be determined:.

$$\widetilde{M_\alpha}\sqrt{\frac{2}{\kappa-1}\left[\frac{\left(\frac{\tilde{U}_1}{U_{\max}}\right)^2}{\left[1-\left(\frac{\tilde{U}_1}{U_{\max}}\right)^2\right]}\right]}$$

Thus a Mach number of the flow is to be assigned to each statement of an area-averaged velocity standardized with the maximum velocity.

All quantities which are stated in the above equations can also be written as functions of the Mach number $\widetilde{M_\alpha}$ which in turn is to be considered as an area-averaged flow quantity describing the course of the flow along the x_1 – axis.

For the derivation of the dependency of the pressure, the density and the temperature from the Mach number of the flow, equation is written as follows:

$$c_P\tilde{T} + \frac{1}{2}\tilde{U}_1^2 = c_P T_H$$

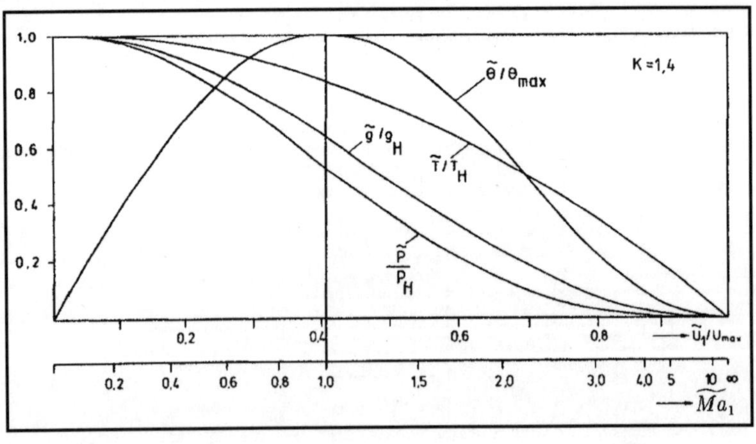

Fig: Course of the Pressure, the Density, the Temperature and the mass-flow Velocity in Pressure-Compensating Flows

By division with $c_p\tilde{T}$ one obtains:

$$\frac{T_H}{\tilde{H}}=1+\frac{\tilde{U}_1^2}{2c_p\tilde{T}}\frac{\kappa R}{\kappa R}=1+\frac{\kappa-1}{2}\widetilde{M_{\alpha_1}}^{2}$$

or for the reciprocal:

$$\frac{\tilde{T}}{T_H}=\frac{2}{2+(\kappa-1)\widetilde{M\alpha_1^2}}$$

This equation makes it clear that a relation is given between the area-averaged temperature along the x_1 – axis and the Mach number existing at the same point of the flow.

With this for each x_1 – point the temperature can be computed, when the reservoir state is given and the Mach number of the flow known.

Taking into account the adiabatic equation, for the relation of pressure and reservoir pressure results:

$$\frac{\tilde{P}}{P_H}=\left(\frac{\tilde{T}}{T_H}\right)^{\frac{\kappa}{\kappa-1}}=\left[\frac{2}{2+(\kappa-1)\widetilde{M\alpha_1^2}}\right]^{\frac{\kappa}{\kappa-1}}$$

and

$$\frac{\tilde{\rho}}{P_H}=\left(\frac{\tilde{T}}{T_H}\right)^{\frac{\kappa}{\kappa-1}}=\left[\frac{2}{2+(\kappa-1)\widetilde{M\alpha_1^2}}\right]^{\frac{\kappa}{\kappa-1}}$$

Die Fig. 9.11 also contains the course of the mass-flow density $\theta = \dot{m}/F = \tilde{\rho}\tilde{U}_1$,., *i.e.* the statement of the mass flowing per area and time unit through a flow cross-section. The course of this quantity can be written as follows, using the relations for $\tilde{U}_1$ and $\tilde{\rho}$:

$$\tilde{\rho}_1\tilde{U}_1=\rho_H\left[1-\left(\frac{\tilde{U}_1}{U_{\max}}\right)^2\right]^{\frac{\kappa}{\kappa-1}}\cdot\tilde{U}_1^2$$

or for the standardized mass-flow density:

$$\frac{\tilde{\rho}_1\tilde{U}_1}{\rho_H U_{\max}}=\frac{\tilde{U}_1}{U_{\max}}\left[1-\left(\frac{\tilde{U}_1}{U_{\max}}\right)^2\right]^{\frac{\kappa}{\kappa-1}}$$

The relation indicated above for the mass-flow density makes it clear, that for $U_1 = 0$, $\theta = 0$ is achieved. The mass-flow density, however, assumes the value zero also for $U_1 = U_{max}$ as with setting the maximally possible velocity the density of the fluid also contained in the mass-flow density has dropped to $\tilde{\rho} = 0$ Between these two minimal values the mass-flow density has to traverse a maximum which can be computed by differentiation of the above functions and by setting the derivation to zero. The value obtained by solving the resulting equation has to be inserted for ($\tilde{U_1}$ /U_{max}) in the above equation for the mass-flow density in order to achieve the maximal value. It is computed:

$$\theta_{max} = \rho_H \cdot U_{max} \cdot \sqrt{\frac{\kappa - 1}{\kappa + 1}} \left(\frac{2}{\kappa + 1} \right)^{\frac{1}{\kappa - 1}}$$

where for the velocity value it is obtained:

$$\frac{\tilde{U_1}}{U_{max}} = \sqrt{\frac{\kappa - 1}{\kappa + 1}} \quad \text{for } \theta = \theta_{max}$$

With this the mass-flow density standardized with the maximal value can be written as follows:

$$\frac{\theta}{\theta_{max}} = \sqrt{\frac{\kappa - 1}{\kappa + 1}} \cdot \frac{\tilde{U_1}}{U_{max}} \left[\frac{\kappa + 1}{2} \left(1 - \frac{\tilde{U}_1^2}{U_{max}} \right) \right]^{\frac{1}{\kappa - 1}}$$

The course of this quantity with $\tilde{U_1} / U_{max}$ is also represented in Fig. 9.10

The significance of the maximum of the mass-flow density for the course of compensating flows is dealt with more in detail further down. Its appearance prevents the steady increase of the mass flow with the increase of the pressure difference between pressure reservoirs when the compensating flow takes place via steadily converging nozzles.

A representation of the compensating flows through converging nozzles often regarded to be more simple is achieved by relating the quantities designating the flow to the corresponding quantities of the "critical state", which is designated by Mα = 1 To this state corresponds not only a certain Mach number, *i.e.* $\widetilde{M\alpha_1} = 1$, but also certain values of the thermo-dynamic state quantities: These can be determined from the equations by setting $\widetilde{M\alpha_1} = 1$ From this result the following values for thermo-dynamic state quantities of the fluid in critical state, *i.e.* for $\widetilde{M\alpha_1} = 1$:

$$\frac{\tilde{P}^*}{P_H} = \left(\frac{2}{\kappa+1}\right)^{\frac{\kappa}{\kappa-1}}$$

$$\frac{\tilde{P}^*}{\rho_H} = \left(\frac{2}{\kappa+1}\right)^{\frac{\kappa}{\kappa-1}}$$

$$\frac{\tilde{T}^*}{T_H} = \frac{2}{\kappa+1}$$

With these equations the pressure, density and temperature of a flowing medium can be determined in that cross-section of a converging nozzle in which the sound velocity occurs.

According to the considerations carried out a minimum of the cross-section has to exist at this point.

As at this point the Mach number assumes the value $\widetilde{M\alpha_1} = 1$ the equation can be written as follows:

$$\frac{\tilde{U}_1^{*2}}{U^2_{\max}} = \frac{\kappa-1}{2}\left(\frac{\tilde{T}}{T_H}\right) = \frac{\kappa-1}{\kappa+1}$$

When comparing the values for ($\tilde{U}_1/U_{\max}$) of the relations, one finds that they are identical, *i.e.* the maximum mass-flow density can only occur in the narrowest cross-section of a nozzle, where the sound velocity then also takes place/sets in.

In accordance with the above-stated derivations of the basic equations for pressure-compensating flows between large reservoirs the flow shall be discussed which occurs in a steadily converging nozzle.

The considerations shall be carried out in such a way that the mass flow is computed which results when a certain pressure relation (P_N/P_H) between the reservoirs comes about. Here two pressure ranges are of interest:

$\frac{P_N}{P_H} > \frac{P^*}{P_H}$	The relation of the reservoir pressures is larger than the critical pressure relation
$\frac{P_N}{P_H} < \frac{P^*}{P_H}$	The relation of the reservoir pressures is smaller than the critical pressure relation

When the pressure relation is larger than the critical value, a steady decrease of the relation of the reservoir pressures leads to a steady increase of the mass-flow density.

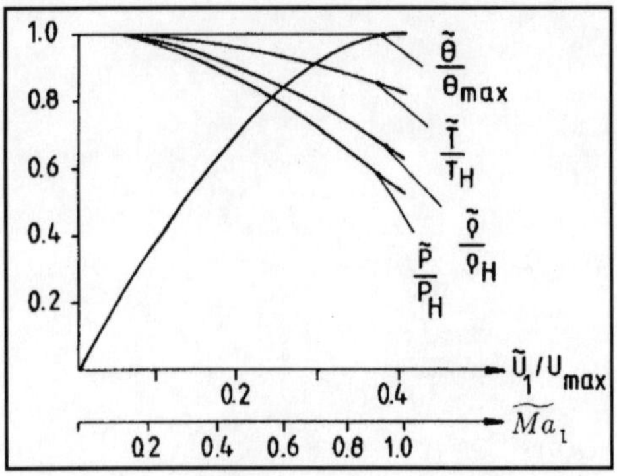

Fig: Course of the Pressure, the Density, the Temperature and the Massflow Density for Converging Nozzles

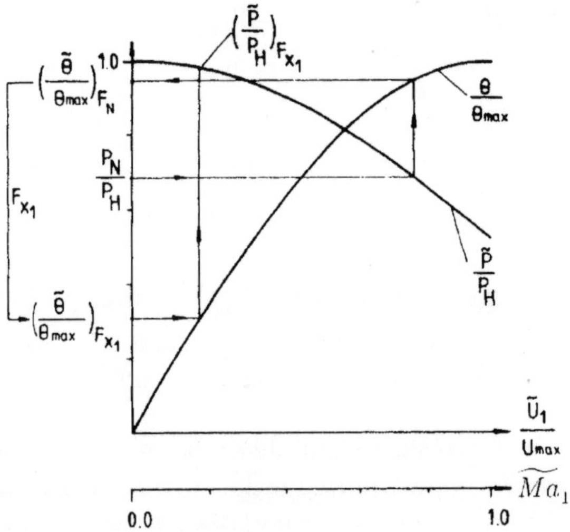

Fig: Determining the Pressure Distribution along the Nozzle axis for $(P_N/P_H) > (P^*_N/P_H)$

On the assumption, that in the narrowest cross-section of the steadily converging nozzle the pressure of the low-pressure reservoir sets in, the pressure relation (P_N/P_H). can be determined with the known values P_N. and P_H–. Via the same the mass-flow density in this cross-section can be determined in the below-stated manner and thus also the total mass flowing through the nozzle:

$$\dot{m}_H = F_H \tilde{\Theta}_H = F_H \left(\tilde{\rho}\tilde{U}_1\right)_H$$

For reasons of continuity this total mass flow is constant in all cross-section planes of the nozzle, so that it holds:

$$\dot{m}_H = \dot{m} \quad \text{d.h.} \ F_N \tilde{\theta}_N = F_{x_1} \tilde{\theta}_{x1}$$

Starting from the assumption that the specified/given distribution of the cross-section area of the nozzle along the x_1 – axis is known, then the massflow density distribution along the x_1 – axis can be determined. Via the same can then be computed, the pressure distribution along the nozzle, or the resulting distributions of the density and the temperature, but also of the Mach number and the flow velocity.

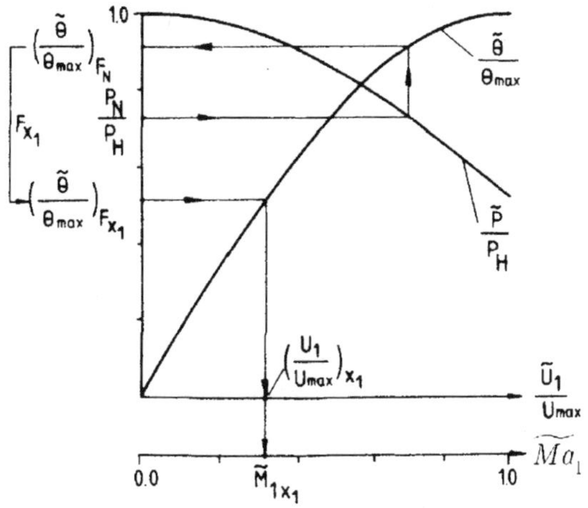

Fig: Determining the Mach number and the Velocity Distribution along a Converging nozzle for $(P_N/P_H) > (P^*_N/P_H)$

The way of proceeding in determining the pressure distribution along the nozzle, indicated in the above figure, can be transferred analogously also to defining the density distribution and the temperature distribution. For determining the distribution of the Mach number and the velocity, the way indicated in Fig. 9.12 holds.

From the above considerations follows that the velocity $(U_1)_N$ in the entrance cross-section of the nozzle is finite and that there the mass-flow density

$$\tilde{\theta}_H = \frac{F_N}{F_H}\left(\tilde{\rho}\tilde{U}_1\right)_N$$

is present. With this it is also said that in this cross-section a pressure, a density and a temperature are reigning which do not correspond to the values in the high-pressure reservoir. It is necessary to take this always into consideration when computing compensating flows through nozzles. The quantities designating the flows that exist at the nozzle entrance are to be determined via the above diagrams from the mass-flow density computed for the entrance cross-section.

When carrying out the above computations for determining the flow quantities and the thermo-dynamic quantities, it proves that with a decrease of the pressure relation (P_N/P_H) an increase of the mass-flow density in each cross-section of the nozzle is connected, as long as the pressure relation is larger than the critical value. When the critical value itself is reached, *i.e.*:

$$\frac{P_N}{P_H} = \left(\frac{2}{\kappa+1}\right)^{\frac{\kappa}{\kappa-1}} = \frac{P^*}{P_H}$$

This value cannot be exceeded in the case of a further decrease of the pressure relation

(P_N/P_H) *i.e.* for all pressure relations smaller than the critical value:

$$\frac{P_N}{P_H} < \frac{P^*}{P_H} = \left(\frac{2}{\kappa+1}\right)^{\frac{\kappa}{\kappa-1}}$$

in the steadily converging nozzle a flow comes about which is identical for all pressure relations. At the exit cross-section of the nozzle, *i.e.* in the entrance cross-section to the low-pressure reservoir, the pressure P_N does not come about any more. In this cross-section the maximum mass-flow density rather is reached:

$$\tilde{\theta}_{\max} = \rho_H \sqrt{\frac{2\kappa}{\kappa-1}\frac{P_H}{\rho_H}} = \sqrt{\frac{2\kappa}{\kappa-1}P_H\rho_H}$$

or

$$\tilde{\theta}_{\max} = \rho_H \sqrt{2c_P T_H}$$

The total mass flow thus is computed:

$$\dot{m} = \dot{m}_{\max} = F_N \tilde{\theta}_{\max}$$

Starting again from the assumption that the nozzle form is known, then the mass-flow distribution existing along the axis can be computed via the continuity equation. When this distribution is known, the corresponding distributions of the pressure, the density, the temperature, the Mach number and the flow velocity can be determined. Of importance is that for all pressure relations (P_N/P_H), that are equal or smaller than the critical relation, one and the same flow comes about in the nozzle. In the exit cross-section of the nozzle for an area-averaged pressure exists which is larger than the pressure P_N. existing in the low-pressure reservoir.

$$\frac{P_N}{P_H} < \frac{\tilde{P}^*}{P_H} = \left(\frac{2}{\kappa+1}\right)^{\frac{\kappa}{\kappa-1}}$$

The pressure compensation takes place via fluid flows that form in the open-jet flow, stretching from the nozzle tip to the interior to the low-pressure reservoir.

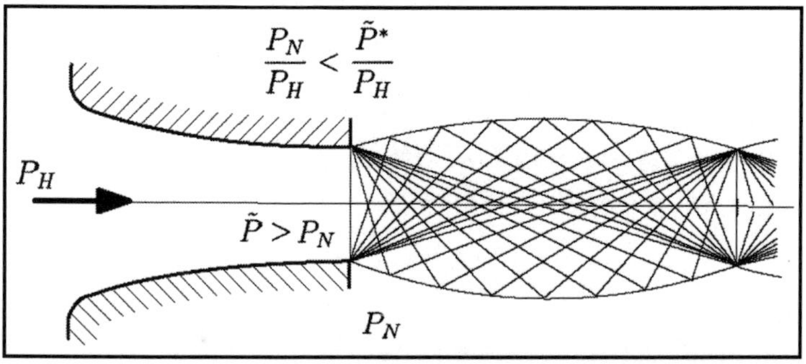

Fig: Pressure Compensation at the Nozzle exit Via Density Impacts

Finally attention is drawn to important facts of the case which shall serve the comprehension of the occurring compensating flow. The above represen tations started from the state often existing in practice that compensating flows are controlled via pressure differences between reservoirs. This means that it was assumed that P_H, ρ_H or T_H are known and constant and that they have an influence on how the flow forms. In the low-pressure reservoir it was only assumed that P_N is given and can be "forced upon" the flow in the narrowest cross-section of the nozzle (for (P_N/P_H) larger than the critical value (P^*/P_H)). The density of the flowing gas coming about for these conditions in the exit cross-section of the nozzle or the occurring temperature are not identical with the corresponding values of the fluid in the low-pressure reservoir.

A compensation of these values and the corresponding values of the lowpressure reservoir takes place in the open-jet flow following the nozzle flow. For pressure conditions:

$$\frac{P_N}{P_H} < \frac{P^*}{P_H} = \left(\frac{2}{\kappa+1}\right)^{\frac{\kappa}{\kappa-1}}$$

the compensation takes place between the pressure in the nozzle-exit crosssection and the pressure in the low-pressure reservoir, likewise in the open-jet flow following the nozzle flow.

HYDROSTATIC PRESSURE

Considering a small cube of water at rest below a free surface, the weight of the water above must be balanced by a pressure in this small cube. For an infinitely small cube the stress is the same in all directions and water weight or equivalent pressure can be expressed as,

$$P = \rho g h$$

where, using SI units,

P is the hydrostatic pressure (in pascals);
ρ is the water density (in kilograms per cubic metre);
g is gravitational acceleration (in metres per second squared);
h is the height of fluid above (in metres).

ATMOSPHERIC PRESSURE

The ideal gas law predicts that, for a gas of constant temperature, T, its density, ρ, will vary with height, h, as:

$$\rho(\mathrm{h}) = \rho(0)\mathrm{e}^{-\mathrm{Mgh/RT}}$$

where:

g = The acceleration due to gravity
T = Temperature
R = Ideal gas constant
M = Molar mass
ρ = Density
h = height.

Buoyancy

A solid body immersed in a fluid will have an upward buoyant force acting on it equal to the weight of displaced fluid. This is due to the hydrostatic pressure in the fluid. In the case of a container ship, for instance, its weight force is balanced by a buoyant force from the displaced water, allowing it to float. If more cargo is loaded onto the ship, it would sit lower in the water - displacing more water and thus receive a higher buoyant force to balance the increased weight force. Discovery of the principle of buoyancy is attributed to Archimedes.

Stability

A floating object is stable if it tends to restore itself to an equilibrium position after a small displacement. For example, floating objects will generally have vertical stability, as if the object is pushed down slightly, this will create a greater buoyant force, which, unbalanced against the weight force will push the object back up.

Rotational stability is of great importance to floating vessels. Given a small angular displacement, the vessel may return to its original position (stable), move away from its original position (unstable), or remain where it is (neutral).

Rotational stability depends on the relative lines of action of forces on an object. The upward buoyant force on an object acts through the centre of buoyancy, being the centroid of the displaced volume of fluid. The weight force on the object acts through its centre of gravity. An object will be stable if an angular displacement moves the line of action of these forces to set up a 'righting moment'.

LIQUIDS-FLUIDS WITH FREE SURFACES

Liquids can have free surfaces at which they interface with gases, or with a vacuum. In general, the lack of the ability to sustain a shear stress entails that free surfaces rapidly adjust towards an equilibrium. However, on small length scales, there is an important balancing force from surface tension.

NOZZLE FLOW METER

The nozzle as shown is practically a venturi with the diverging part removed. The basic equations used are the same as for the venturi meter.

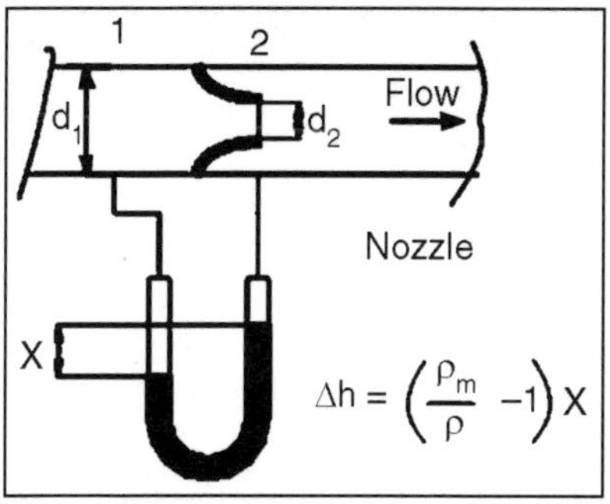

Fig. Design and Performance Parameters of Nozzles

The friction losses are slightly larger than for the venturi but this is offset by the lower cost of the unit. The fact that the manometer connections cannot be located in the ideal positions for measuring the required piezometric pressures is allowed for in selection of the coefficient of discharge factor C_d..

ORIFICE FLOW METER

The simplest and cheapest method of measuring the flow using the bernoulli equation is the sharp edged orifice as shown below.

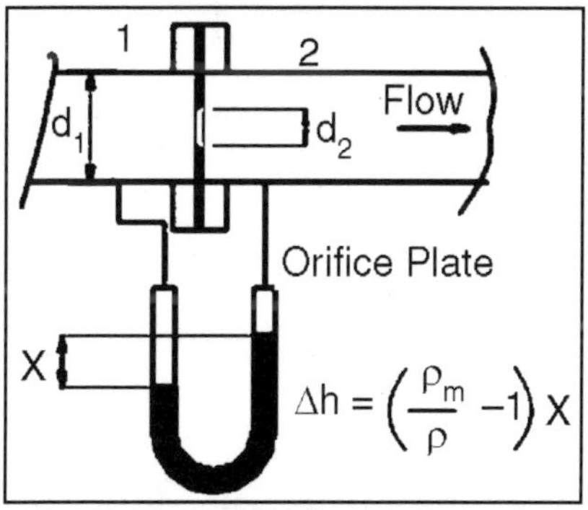

Fig. The Fluid Flow Pattern.

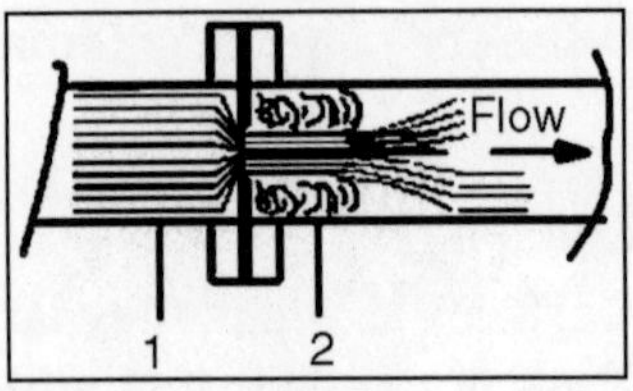

Fig. Application of Bernoulli's equation.

$$\frac{u_2^2}{2g}=\frac{p_1}{\rho g}-\frac{p_2}{\rho g}+\frac{u_1^2}{2g}=\Delta h+\frac{u_1^2}{2g}$$

Introducing for real fluid the coefficient of velocity C_v,

$$u_2 = c_v \left[2g\left(\Delta h+\frac{u_1^2}{2g}\right)\right]^{1/2}$$

Now $u_1 = Q/A_1$ and $u_2 = Q/A_c$ where A_c = The area of the vena-contracta which is the reduced area of the fluid after leaving the orifice hole. ($A_c = C_c A_2$).

A_2 = the area of the orifice and C_c is the Coefficient of contraction.

Finally C_d = Coefficient of discharge = $C_c.C_v$

Using these factors a relationship for Q can be developed from the above equation,

$$Q=\frac{C_v\, C_c\, A_2\, A_1}{(A_1^2-C_v^2\, C_c^2\, A_2^2)^{1/2}}(2g\,\Delta h)^{1/2}=\frac{C_d\, A_2\,(2g\,\Delta h)^{1/2}}{[1-C_d(A_2/A_1)^2]^{1/2}}$$

To arrive at a final equation a overall discharge coefficient C is introduced.

$$C=C_d\left(\frac{1-(A_2/A_1)^2}{[1-C_d^2(A_2/A_1)^2]^{1/2}}\right)$$

Now letting $\beta = d_2/d_1$ that is $\beta^2 = A_2/A_1$. The equation for flow through an orifice becomes,

$$Q=\frac{C\,A_2(2g\,\Delta h)}{(1-\beta^4)^{1/2}}$$

This equation is very similar to the equation provided in BS EN ISO 5167:2 except that an expansion coefficient (ε)is introduced to cater for the measurement of compressible fluids.

The equation provided in the standards is,

$$Q_m=\frac{C}{\sqrt{(1-\beta^4)}}\varepsilon\frac{\pi}{4}d^2\sqrt{2\Delta p\rho_1} \quad \ldots\ldots Q_m=Q\rho_1$$

Values of the discharge coefficient C are provided in BS EN ISO 5167:2 for the different meter tapping arrangements, for different values of β against Reynold number ranges.

Small table showing C values for different Reynold numbers and β values. Detailed tables are provided in BS EN ISO 5167:2

b	Re		
	5×10^3	1×10^5	$1 \times 10^{8<}$
0.25	0,6102	0,6025	0,6013
0.5	0,6284	0,6082	0,6036
0.75	0,6732	0,6171	0,6025

FLOW CONDITIONING

The accuracy of the flow measuring devices is very much affected by uniformity of the approaching fluid flow. Therefore ideally there should be a straight length of piping before the flow measuring device. It is generally accepted that for accurate flow readings there should be 50 pipe diameters of straight piping before the metering device following any pipe bend, valve, tee, reducer etc. The relevant standard provides a range of recommended minimum straight lengths depending on the nearest upstream fittings varying from 5 to 44 lengths.

This length can be reduced if flow straighteners or flow conditioning devices systems are used upstream of the flow measuring device. A flow straightener is designed to remove swirl from the flowing fluid. A flow conditioner is a device which removes swirl and also redistributes the velocity profile to produce near ideal metering conditions.

LOSSES RESULTING FROM FLOW METERING DEVICES

The orifice plate and, to a lesser extent the nozzle has significant kinetic energy losses downstream of the metering device as the locally generated kinetic energy is dissipated. The figurc below illustrates the extent of these losses.

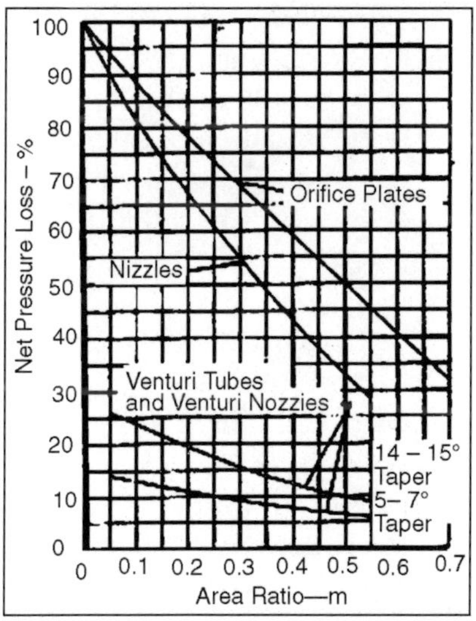

The Madden Orifice Meter gives the boiler operator precise repeatable control of the continuous blowdown flow rate. 17 different hole sizes are drilled in the stainless steel orifice plate. Each hole is precision drilled to a diameter that will produce the flow shown on he flow chart. By selecting a larger hole the continuous blowdown flow is increased, or by selecting a smaller hole number on the index band the blowdown is reduced. Four different orifice plates (1A, 2A, 3A, and 4A) are available to provide a blowdown flow range suitable for the boiler operator to maintain the target boiler water quality level.

The boiler must have a continuous blowdown skimmer pipe installed inside the boiler drum, with the opening for the pipe within 6" of the normal water level in the boiler drum.

The Madden Orifice Meter may be mounted near the boiler or near a continuous blowdown heat recovery system. The constant blowdown flow provided by the Madden Orifice Meter makes it an ideal device for use in conjunction with a blowdown heat recovery system, see Madden Blowdown heat recovery system for details.

The Madden Orifice Meter has been in production for almost 50 years. The units are generally used on industrial boilers from 200 HP up to over 100,000 PPH steam production, operating at pressures from 50 psi up to 650 psi. One unit is required for each boiler. The meter can be utilized as the sole flow control device or pipe in parallel with a conductivity sensing blowdown control.

The units are also used with non-fired boilers and reboilers used in refineries. The Madden Orifice Meter is guaranteed for up to 10 years against damage from cutting, wire drawing, or other distortion in the orifice plate.

Part Number	Maximum	Description	Weight lbs.	PSI
OM2501A	250	Orifice meter with orifice plate 1A installed	64	
OM2502A	250	Orifice meter with orifice plate 2A installed	64	
OM2503A	250	Orifice meter with orifice plate 3A installed	64	
OM2504A	250	Orifice meter with orifice plate 4A installed	64	
OM6501A	650	Orifice meter with orifice plate 1A installed	80	
OM6502A	650	Orifice meter with orifice plate 2A installed	80	
OM6503A	650	Orifice meter with orifice plate 3A installed	80	
OM6504A	650	Orifice meter with orifice plate 4A installed	80	

VENTURI METER

In comparing the venturi meter with the orifice meter, both the cost of installation and the cost of operation must be considered.

- The orifice plate can easily be changed to accomodate widely different flow rates, whereas the throat diameter of a venturi is fixed, so that its range of flow rates is circumscribed by the practical limits of Dp.
- The orifice meter has a large permanent loss of pressure because of the presence of eddies on the downstream side of the orifice-plate; the shape of the venturi meter prevents the formation os these eddies and greatly reduces the permanent loss.
- The orifice is cheap and easy to install. The venturi meter is expensive, as it must be carefully proportioned and fabricated. A home made orifice is often entirely satisfactory, whereas a venturi meter is practically always purchased from an instrument dealer.
- On the other hand, the head lost in the orifice for the same conditions as in the venturi is many times greater. The power lost is proportionally greater, and, when an orifice is inserted in a line carrying fluid continuously over long periods of time, the cost of the power may be out of all proportion to the saving in first cost. Orifices are therefore best used for testing purposes or other cases where the power lost is not a factor, as in steam lines.
- However, in spite of considerations of power loss, orifices are widely used, partly because of their greater flexibility, because installing a new orifice plate with a different opening is a simpler matter. The venturi meter can not be so altered. Venturi meters are used only for permanent installations.
- It should be noted that for a given pipe diameter and a given diameter of orifice opening or venturi throat, the reading of the venturi meter for a given velocity is to the reading of the orifice as $(0.61/0.98)^2$, or 1:2.58.(*i.e.* orifice meter will show higher manometer reading for a given velocity than venturi meter).

BAND METER

Amateur Radio Bands

Just as many of us have a favorite fast food that we come back to time after time Radio Amateurs tend to have a favorite band that they use more frequently than other bands. The question of what band to use may be secondary to that favorite band.

But the first question should be what band(s) am I licensed to use? From the beginning of the licensing process you will find out what bands your license covers.

Next it is a matter of having the right equipment for the bands you want to use. As an entry level license generally covers VHF(2m) and UHF(70cm) and many beginning hams will use one of both of these bands. Also hand-held single-band or dual-band radios for these bands are affordable and don't require a sophisticated antenna or power source. Most metropolitan areas also have amateur repeaters to extend the coverage when using VHF or UHF radios.

Band Restrictions

In addition to deciding what band to use there may also be restrictions within the band depending on your level of license. Canada and the U.S. each have slightly different band allocations which amateurs in each country must follow. See the sidebar for links to web sites where the band plans are defined.

Also the world is divided into 3 ITU regions each of which have their own band plan. These plans all have a great deal of overlap so you can usually talk to anyone anywhere providing you get onto a frequency acceptable to both.

Popular Bands

So what are the bands that most Radio Amateurs use? The table below shows bands that are relatively common and how and when they are in use. Again your license will determine which bands and what portions you are eligible to use.

	Band (meter)	MHz	Use*
HF	160	1.8 – 2.0	night
	80	3.5 – 4.0	night and local day
	40	7.0 – 7.3	night and local day
	30	10.1 - 10.15	CW and digital
	20	14.0 - 14.350	world wide day and night
	17	18.068 - 18.168	world wide day and night
	15	21.0 - 21.450	primarily a daytime band
	12	24.890 - 24.990	primarily a daytime band
	10	28.0 - 29.70	daytime during sunspot highs
VHF	6	50 - 54	local to world-wide
	2	144 - 148	local and medium distance
UHF	70 cm	430 - 440	local

* It should be noted that band conditions vary for many reasons and thus all of these bands can at times take on the characteristics of others. See the section on Propagation. This table should be considered a general guideline.

HF Bands

For amateurs who have an interest in long distance communication the HF (high frequency) bands will be of great interest. These bands offer propagation to all parts of the world at some time during the day, night or season.

Do you want to talk to Japan from the east coast or Europe from the west? There will often be a time and a band where this communication is possible.

During a sunspot high cycle conditions are best for the higher HF frequencies and during a low the low frequencies are often in demand with all kinds of variation in between. Magazines such as QST and CQ Amateur Radio publish charts monthly that predict the best propagation to different areas of the world. These are just best guesses and will help you to choose your times and bands but there is no substitute for getting on the air.

6

Vortex Motion and Dynamics

VORTEX

Vortex created by the passage of an aircraft wing, revealed by coloured smoke. A vortex (pl. *vortices*) is a spinning, often turbulent, flow (or any spiral motion) with closed streamlines. The shape of media or mass swirling rapidly around a center forms a vortex. It flows in a circular motion.

Dynamics

A vortex can be any circular or rotary flow that possesses vorticity. Vorticity is a mathematical concept used in fluid dynamics. It can be related to the amount of "circulation" or "rotation" in a fluid. In fluid dynamics, vorticity is the circulation per unit area at a point in the flow field. It is a vector quantity, whose direction is (roughly speaking) along the axis of the swirl. Also in fluid dynamics, the movement of a fluid can be said to be vortical if the fluid moves around in a circle, or in a helix, or if it tends to spin around some axis. Such motion can also be called solenoidal. In the atmospheric sciences, vorticity is a property that characterizes large-scale rotation of air masses. Since the atmospheric circulation is nearly horizontal, the (3 dimensional) vorticity is nearly vertical, and it is common to use the vertical component as a scalar vorticity. Mathematically, it is defined as,

$$\vec{w} = \nabla \times \vec{u}$$

where $\vec{u}$ is the *fluid* velocity.

Two Types of Vortex

In fluid mechanics, a distinction is often made between two limiting vortex cases. One is called the free (irrotational) vortex, and the other is the forced (rotational) vortex. These are considered as below:

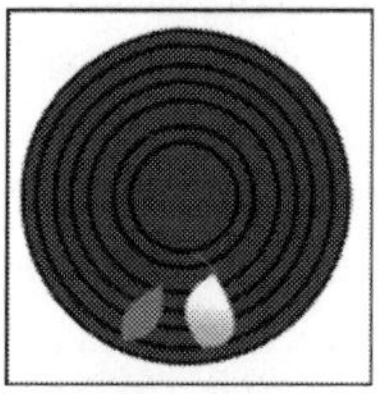

Fig. Two Autumn Leaves in a Counter-clockwise Vortex (Reference Position).

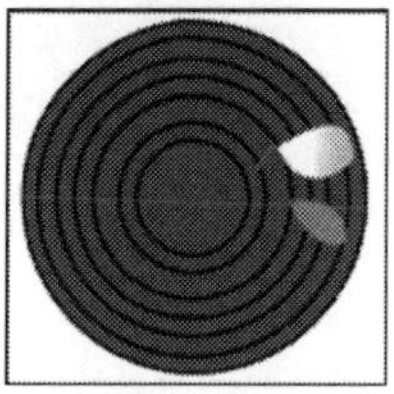

Fig. Two Autumn Leaves in a Rotational Vortex Rotate with the Counter-clockwise Flow.

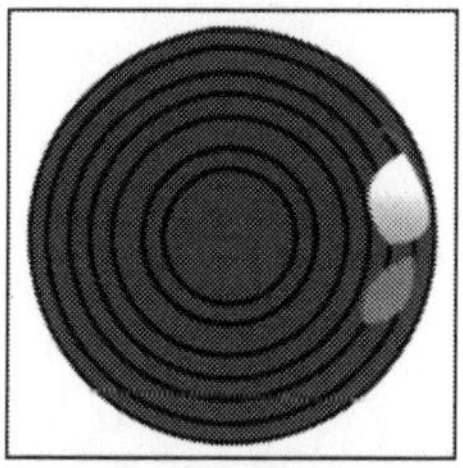

Fig. Two Autumn Leaves in an Irrotational Vortex Preserve their Original Orientation While Moving Counter-clockwise.

FREE (IRROTATIONAL) VORTEX

When fluid is drawn down a plug-hole, one can observe the phenomenon of a free vortex. The tangential velocity v varies inversely as the distance r from the center of rotation, so the angular momentum, rv, is constant; the vorticity is zero everywhere (except for a singularity at the center-line) and the circulation about a contour containing $r=0$ has the same value everywhere. The free surface (if present) dips sharply (as r^{-2}) as the center line is approached.

The tangential velocity is given by:

$$v_\theta = \frac{\Gamma}{2\pi r}$$

where Γ is the circulation and r is the radial distance from the center of the vortex. In non-technical lingo, the circular streamlines towards the center can sweep out a given angle faster than the outer streamlines. The speed along the circular path of flow is held constant or decreases as you move out from the center. At the same time the inner streamlines have a shorter distance to travel to complete a ring. If you were running a race on a circular track would you rather be on the inside or outside, assuming the goal was to complete a circle? Imagine a leaf floating in a free vortex. The leaf's tip points to the center and the blade straddles multiple streamlines. The outer flow is slow in terms of angle traversed and it exerts a backwards tug on the base of the leaf while the faster inner flow pulls the tip forwards. The drag force opposes rotation of the leaf as it moves around the circle.

In a forced vortex the fluid essentially rotates as a solid body (there is no shear). The motion can be realised by placing a dish of fluid on a turntable rotating at T radians/sec; the fluid has vorticity of 2 T everywhere, and the free surface (if present) is a parabola.

The tangential velocity is given by:

$$v_\theta = wr$$

where ω is the angular velocity and r is the radial distance from the center of the vortex.

OBSERVATIONS

A vortex can be seen in the spiraling motion of air or liquid around a center of rotation. Circular current of water of conflicting tides form vortex shapes. Turbulent flow makes many vortices. A good example of a vortex is the atmospheric phenomenon of a whirlwind or a tornado or dust devil. This whirling air mass mostly takes the form of a helix, column, or spiral. Tornadoes develop from severe thunderstorms, usually spawned from squall lines and supercell thunderstorms, though they sometimes happen as a result of a hurricane.

A mesovortex is on the scale of a few miles (smaller than a hurricane but larger than a tornado). On a much smaller scale, a vortex is usually formed as water goes down a drain, as in a sink or a toilet. This occurs in water as the revolving mass forms a whirlpool. This whirlpool is caused by water flowing out of a small opening in the bottom of a basin or reservoir. This swirling flow structure within a region of fluid flow opens downward from the water surface.

MOTION OF VACANCIES IN A PINNED VORTEX LATTICE

Physical arguments are presented to show that the Hall anomaly is an effect of the vortex many-body correlation rather than that of an individual vortex. Quantitatively, the characteristic energy scale in the problem, the vortex vacancy formation energy, is obtained for thin films. At low temperatures a scaling relation between the Hall and longitudinal resistivities is found, with the power depending on sample details. Near the superconducting transition temperature and for small magnetic fields the Hall conductivity is found to be proportional to the inverse of the magnetic field and to the quadratic of the difference between the measured and the transition temperatures. The ubiquitous occurrence of the Hall anomaly in the mixed state of both conventional and oxide superconductors, the sign change of the Hall resistivity below the superconducting transition temperature and the smallness of the Hall angle, has defied a consistent explanation so far. A straightforward application of the Magnus force cannot explain this phenomenon. This failure leads to the frustration of questioning the Magnus force as the only transverse force in the vortex dynamic equation. The transverse force has been subsequently modified into various forms. Following the general properties of a superconductor, on the other hand, recent theoretical work on vortex dynamics has shown that there is no transverse force other than the Magnus force, a result of the topological constraint and the momentum conservation. One can further show that the Magnus force is equivalent to the spectral flow process. An apparent conflict between the theoretical reasonings and the experimental measurements naturally arises.

In this letter we attempt to solve this puzzle by showing that the Hall anomaly can be understood based on the vortex vacancy motion in a pinned vortex lattice, and emphasize that the anomaly is a property of the vortex many-body correlation rather than that of an individual vortex. We will demonstrate that vacancies can have the lowest energy scale, and that they dominate the thermal activation contributions to the vortex motion at low temperatures. The present vacancy model for the Hall anomaly is also consistent with other measurement such as the Nernst effect. In the following we present our arguments leading to the model, and discuss its predictions and validity conditions. For simplicity, we will consider an isotropic s-pairing superconductor with one type of charge carrier in two dimension. In this situation vortices (or straight vortex lines) can be viewed as point particles.

The vortex dynamic equation for a jth vortex of unit length in the sample takes the form of the Langevin equation identical to that of a charged particle in the presence of a magnetic field:

$$m\ddot{r}_j = q\frac{\rho_s}{2}h\left(v_{ss} - \dot{r}_j\right)\times\hat{z} - \eta\dot{r}_j + F_p + f$$

with an effective unit length mass m, a pinning force F_p, a vortex viscosity η, and a fluctuating force f. In equation (1) $q = \pm 1$ is the vorticity, h the Planck constant, ρ_s the superfluid electron number density at temperature T, and $\hat{z}$ the unit vector in z-direction. If there is a temperature gradient, the thermal force $F_\tau = -s_\phi \nabla T$ should be added in at the right hand of equation (1), with s_ϕ

the entropy carried by a vortex. The term associated with the total superconducting electron velocity $v_{s,s} = v_s + v_{s,in}$ and the vortex velocity $\dot{r}$ at the right side of equation (1) is the Magnus force. Although $v_{s,s}$ is due to all other vortices, here we split it into two parts, with v_s corresponding to the rearrangement of vortices due to the externally applied supercurrent and $v_{s,in}$ accounting for the rest contribution describing the vortex interaction without external current. In the following we will assume v_s is small such that this splitting is valid.

It is evident that in the mixed state of any real superconductor the many-body correlation between vortices and the pinning effect cannot be ignored. The competition between them is the source of the rich static and dynamical properties of flux phases. We will take the Abrikosov lattice as the known manifestation of the many-body correlation for the starting point to advance our arguments. The vortex pinning is also important in our reasoning, though several quantitative results obtained below do not explicitly depend on it. If there were no pinning for vortices, the whole vortex lattice would move together under the influence of an externally applied current in the same manner as that of independent vortices.

Hence, one would get the same sign of the Hall resistivity in both superconducting and normal states. In the presence of pinning centres in the sample as well as the edge pinning, the vortex lattice will be pinned down. In such a situation the motion of the vortex lattice is made possible by various kinds of defect motions due to thermal fluctuations. We will argue below that at low temperatures the dominant contribution to the motion is due to vortex vacancies, and the Hall anomaly occurs.

For two vortices separated by a distance r, which is less than the effective magnetic screening length,

$$\lambda \perp \lambda_L^2 / d \left(d < \lambda_L \, \lambda_\perp = \lambda_L \text{ if } d > \lambda_L \right)$$

but greater than ξ_0, the interaction potential is,

$$V_o(r) = 2d \left(\Phi_0 / 4\pi\lambda_L \right)^2 \ln \left(r / \xi_0 \right)$$

Here,

$$\lambda_L^2 = m * c^2 / 8\pi\rho_s e^2$$

is the London penetration depth, m* the effective mass of a Cooper pair, ξ_0 the coherence length of the superconductor, and d the thickness of the superconductor film. The energy scale,

$$\epsilon_0 \equiv d \left(\Phi_0 / 4\pi\lambda_L \right)^2$$

sets both the scale for the strength of vortex interaction and the scale for the strength of a strong pinning centre. The energy for a dislocation pair separated by a distance larger than the vortex lattice constant is given by,

$$T << V_d(r) = \left(\epsilon_0 / 2\sqrt{3}\pi \right) \ln \left(r / a_0 \right)$$

with a_0 an order of the the vortex lattice constant. The energy scale $\epsilon_0 / 2\sqrt{3}\pi$ for the dislocation pair here is about ten times smaller than ϵ_0 for the vortex interaction and pinning centres. It is energetically favourable to have dislocation pairs in the lattice. Hence, for temperature $T \ll \epsilon_0$ we can ignore the contribution from the vortices hopping out of pinning centres and the creation of vortex - antivortex pairs. The vortex lattice is then effectively pinned down. Because vacancies and interstitials can be viewed as the smallest dislocation pairs, we immediately have the estimated energy scale for vacancy formation energy ϵ_v as, by putting $r \sim 2a_0$ in $V_d(r)$,

$$\epsilon_v \sim \frac{1}{2\sqrt{3}\pi}\left(\frac{\Phi_0}{4\pi\lambda_L}\right)^2 d.$$

This result is valid for an intermediate magnetic field B: $H_{cL} < B < H_{c2}/2$. For thicker films the thickness in equation will be replaced by a crossover thickness d_c due to the z-direction correlation, whose precise value is a complex and unknown function of various parameters such as the magnetic field, the pinning, the temperature, and anisotropy. In the case d_c is finite, its estimation in the high magnetic field limit is as follows. Ring type vacancy excitations are possible in thicker films. Its energy scale is determined by the smallest ring, which should be the size of the vortex lattice constant. In this case d_c is an order of the vortex lattice constant, $d_c \sim a_0$. However, because of the large anisotropy in the HTcS materials, d_c can be the order of the CuO layer spacing close to the superconducting transition temperature T_{c0}.

It is clear from the above analysis that vacancies and interstitials have the lowest excitation energy scales. We note that the value in equation (2) is consistent with the variational and numerical calculations, with about a factor 2 smaller, and also with the estimation from the dislocation core energy. Using the shear modulus results for $B \le H_{cL}$ and,

$$H_{c2}/2 < B \le H_{c2}$$

we have obtained the corresponding vacancy formation energies as,

$$(2/9\pi)^{L/2}(b_0/\lambda_L)^{3/2} e^{-b0/\lambda L} \epsilon_0$$

with,

$$b_0^2 = \Phi_0 / B$$

and

$$0.7/\sqrt{3}\pi(1 - B/H_{c2})^2 \epsilon_0$$

respectively.

Now we argue that the vacancy formation energy is even lower than that of an interstitial. The experimental observations at low magnetic fields have shown the abundance of vacancies comparing with interstitials. The natural explanation is that the the vacancy formation energy is lower than that of interstitials, therefore by thermal fluctuations vacancies have a higher density.

The theoretical calculations have also confirmed the lower vacancy formation energy. This phenomenon of the vacancy formation energy is lower than that of interstitials has also been observed in other crystalline structures. We conclude that vacancies will dominate thermal fluctuation contributions to resistivities at low enough temperatures for low magnetic fields.

We show next that in the pinned vortex lattice a vacancy behaves as a vortex with a vorticity - q and an interstitial as a vortex with + q, respectively. Let u be the displacement vector at position r, with a point defect, vacancy or interstitial, at r_0. According to equation, the transverse force acting on the defect is, measured from the pinned perfect vortex lattice,

$$F_M^d = q\frac{\rho_s}{2} h \int d^2 r \delta\rho(r)(v_s - \dot{u}) \times \dot{z}$$

Here the vortex density $\delta\rho$ deviated from a perfect lattice is determined by the dilatation,

$$\nabla \cdot u : \delta\rho = \nabla \cdot u / S_0$$

with S_0 the area of a unit cell in the vortex lattice. By definition,

$$\nabla \cdot u = \mp S_0 \delta^2 (r - r_0)$$

with '-' for a missing vortex, a vacancy, and '+' for an extra vortex, an interstitial. Using equations, we have the desired transverse force on the defect as

$$F_M^d = \mp q\frac{\rho_s}{2} h(v_s - \dot{r}_0) \times \dot{z}$$

This is identical to the dynamics of a hole or a particle in a semiconductor in the presence of a magnetic field, with a pinned perfect vortex lattice as a filled valence band and a vacancy in real space as a hole in the energy space. Equation shows that both a vacancy and an interstitial will move along the direction of the applied supercurrent v_s. This implies that vortices defining vacancies move against the direction of v_s, a result of the many-body correlation and pinning.

This leads us to our main conclusion that at low enough temperatures the sign of the Hall resistivity is different from its sign in the normal state because of the dominance of vacancies. Quantitatively, vacancies and interstitials may be considered as independent particles moving in the periodic potential formed by the vortex lattice and a random potential due to the residue effect of pinnings. The potential height of the periodic potential as well as that of the random potential is an order of $\in_v$. Assuming the vacancy (interstitial) density $n_n(n_i)$ in a steady state, the longitudinal resistivity is

$$\rho_{xx} = \frac{h}{2e^2} \sum_{l=vj} \frac{\eta\rho_s h/2}{\eta_l^2 + (\rho_s h/2)^2} \frac{\eta_l}{\rho_s}$$

and the Hall resistivity,

$$\rho_{yx} = \frac{h}{2e^2} \sum_{l=vj} q_l \frac{(\rho_s h/2)^2}{\eta_l^2 + (\rho_s h/2)^2} \frac{\eta_l}{\rho_s}$$

with $q_v = -q$ and $q_i = q$. Here $\eta_{v,i}$ are the effective vacancy and interstitial viscosities, related to their diffusion constants in the periodic potential due to the vortex lattice by the Einstein relation between the diffusion constant and the mobility. It should be pointed out that contributions of other vortex motions to resistivities such as vortex - antivortex pairs, which are omitted here for their smaller activation probabilities, are additional to those of vacancies, and that the including of the normal fluid (quasiparticle) contributions is straightforward.

Under the driving of a temperature gradient, the effective thermal force felt by a vacancy is opposite in sign to the force felt by an interstitial or a vortex in direction but equal in magnitude $F_T^W = +s_\phi \nabla T$. This can be seen by repeating the demonstration from equations. Then the Nernst effect due to vacancies has the same sign as that of vortices or interstitials. Therefore our model gives that in the Hall anomaly regime there is no sign change for the Nernst effect, and furthermore, the Nernst effect is more pronounced because of the additive contributions due to both vacancies and interstitials. This is in agreement with the experimental observations.

Before exploring of consequences of equations we discuss the qualitative implications of the present model. In the above picture, to obtain a maximum contribution of vacancies, we need the vortex lattice to define vacancies and sufficiently strong pinnings to prevent the sliding of the vortex lattice. The existence of a whole lattice structure is nevertheless unnecessary. Sufficiently large local crystalline structures, like lattice domains, will be enough to define vacancies. Therefore vacancy-like excitations in a vortex liquid state can exist, because of the presence of large local orderings. Whether or not this is also true for a vortex glass state depends upon the details. For example, a further lowering of temperature may quench a vortex system into a glass state with no local crystalline structure.

Then vacancies will disappear and the sign of the Hall resistivity will change again. On the other hand, for a fixed temperature if the pinning is too strong, for example, the (random) pinning centre density is much larger than the vortex density, vortices will be individually pinned down and the local lattice structure required for the formations of vacancies and interstitials will be lost. This suggests that the Hall anomaly only exists in a suitable range of pinnings and magnetic fields, that is, for $B_t < |B| < B_u$ with the lower and upper critical fields determined by pinning as well as by temperature.

Now we study the limiting cases of equations. At low temperatures the motions of vacancies and interstitials in the vortex lattice are thermal hopping:

$$\eta_v = \eta_0 e^{a_v \epsilon_v / KaT}$$

and,

$$\eta_i = \eta_0 e^{a_1 \epsilon_v / KaT},$$

with a_v,a_i (presumably $a_v < a_i$) numerical factors of order unity and η_0 insensitive to temperature. In this limit, the vacancy (interstitial) density,

$$\eta_v = \eta_0 e^{-b_v \epsilon_v / KaT} \left(\eta_i = \eta_0 e^{-b_1 \epsilon_v / KaT} \right)$$

with $b_v = 1(b_i > 1)$ for the thermally activated vacancies (interstitials) and a_v $(a_i) = 0$ for the pinning centre induced vacancies (interstitials). In the following we further assume that $\eta_{v,}\eta_i >> \rho_s h / 2$, corresponding to the Hall angle,

$$|\tan\theta| = |\rho_{yx} / \rho_{xx}| << 1$$

common in experiments. Under this assumption, we obtain the Hall angle as,

$$\tan\theta = -q \frac{\rho_s h}{2\eta_0} \frac{e^{-(2a_v + b_v)\epsilon_v / k_B T} - e^{-(2a_1 + b_1)\epsilon_v / k_B T}}{e^{-(a_v + b_v)\epsilon_v / k_B T} + e^{-(a_1 + b_1)\epsilon_v / k_B T}}$$

$$= \begin{cases} -q \dfrac{\rho_s h}{2\eta_0} \dfrac{\gamma}{2} \dfrac{\epsilon_v}{k_B T} & k_B T \geq \epsilon_v \\ -q \dfrac{\rho_s h}{2\eta_0} e^{-a_v \epsilon_v / k_B T} & k_B T < \min\{1, \gamma\} \epsilon_v \end{cases}$$

Here $\gamma = 2a_i + b_i - 2a_v - b_v$. The high temperature limit $k_B T \geq \epsilon_v$ is achieved near the superconducting transition temperature T_{co}, but the thermal creation of a vortex - antivortex pair is still improbable, because the relevant energy scale ϵ_0 is about ten times bigger than ϵ_v. In the low-temperature limit both longitudinal and Hall resistivities vanish exponentially. We obtain a scaling relation between them as,

$$\rho_{yx} = A\rho_x^v x$$

with,

$$A = -q(\rho_s h / 2\eta_0)^{b_v/(a_v + b_v)} (2e^2 \rho s / hn_0)^{a_v/(a_v + b_v)}$$

and the power,

$$v = \frac{2a_v + b_v}{a_v + b_v}$$

varying between 1 and 2, depending on the detail of a sample which determines the numerical factors a_v and b_v. If all vacancies are produced by pinnings, we have $b_v = 0$ and $v = 2$. In this case A is independent of B because n_0 is. In the other limit, if all vacancies are produced by thermal activations, and if $a_v << 1$, we have $a_v = 1$and $v \simeq 1$. In this case A will be independent of B if η_0 is.

Another useful quantity is the Hall conductivity,

$$\sigma_{xy} = \rho_{yx} / \left(\rho_{xx}^2 + \rho_{yx}^2 \right)$$

Under the same assumption of $\eta_v, \eta_i, >> \rho sh/2$ we obtain the Hall conductivity due to vacancies and interstitials, from equations, as,

$$\sigma_{xy} = -q \frac{2e^2}{h} \frac{\rho s}{n_0} \frac{e^{-(2a_v + b_v)\epsilon_v / k_B T} - e^{-(2a_1 + b_1)\epsilon_v / k_B T}}{\left[e^{-(a_v + b_v)\epsilon_v / k_B T} + e^{-(a_1 + b_1)\epsilon_v / k_S T} \right]}$$

$$= \begin{cases} -q \dfrac{2e^2}{h} \dfrac{\rho_s}{n_0} \dfrac{V}{4} \dfrac{\epsilon_v}{k_B T} & k_B T \geq \epsilon_v . \\ -q \dfrac{2e^2}{h} \dfrac{\rho_s}{n_0} e + b_v \epsilon_v / k_s T & k_B T < \min\{1, \gamma\} \epsilon_v . \end{cases}$$

As discussed above, here $0 \leq b_v \leq 1$, and $\gamma \sim 0(1)$. Near the superconducting transition temperature,

$$T_{c0}, \rho_s = \rho_{s0} \left(1 - T / T_{s0} \right)$$

and,

$$\epsilon_v = \epsilon_{v0} \left(1 - T / T_{c0} \right)$$

because of the London penetration depth in equation. We may further assume $n_0 = B / \Phi_0$, with Φ_0 the flux quantum. From equation we obtain,

$$\sigma_{xy} = \alpha_L \frac{\left(1 - T / T_{c0} \right)^2}{B}$$

with,

$$\alpha_L = -q \left(2e^2 / h \right) \rho_{s0} \Phi_0 \gamma \epsilon_{v0} / 4 k_B T_{c0}$$

Taking,

$$\rho_{s0} = 10^{2L} \text{cm}^{-3}, \gamma = 1,$$

and,

$$\epsilon_{v0} / k_B T_{c0} = 50$$

we find,

$$\left| \alpha_L \right| \sim 20 T \mu\Omega^{-1} \text{cm}^{-1}$$

Two points should be noted. (1) A naive accounting of the many-body correlation and pinning may not lead to the sign change in the Hall resistivity: Since vortex interaction terms cancel each other when summing over all vortices, one would like to conclude that there is no many-body correlation effect on the sign of the Hall resistivity. If this claim were correct, the same argument would lead to no sign change for the Hall resistivity in a hole semiconductor, and no such phenomena as the quantum Hall effect. This absence of Hall anomaly is the result of the underestimation of the many-body correlation. (2) There might be a tendency to mix up antivortices and vacancies. As discussed above, the creation energy of an antivortex is about ten times larger than that of a vacancy, which makes it energetically unfavourable. Furthermore, since an antivortex feels the same thermal force as a vortex, $F_T = -s_\phi \nabla T$, it has an opposite sign contribution to that of a vortex for the Nernst effect, in conflict with experiments.

MOTION OF FLUID ELEMENTS

Flow kinematics is a vast field and a comprehensive treatment is beyond the scope of this book, which is meant to give only an introduction into sub-domains of fluid mechanics, among others also into fluid kinematics. To such an introduction belongs the treatment of path lines of particles, *i.e.* the computation of space curves along which marked fluid elements move in a fluid. Further, the computation of sweeping paths shall be treated, *i.e.* the "'marked path"' is to be computed which leaves a fixed injected tracer in a flow field. Both, the computation of path lines and of sweeping paths is of importance for the entire experimental flow mechanics, where it is often tried to get an insight in the process of flows by observations or also by quantitative measurements of the temporal changes of position of "'flow markers"'.

Path Lines of Fluid Elements

When one subdivides at the time $t = 0$ the entire domain of a flow field that is of interest into fluid elements and when one states the space coordinates of the mass centers of gravity of the each element in a coordinate system at the time $t = 0$, one achieves a marked fluid domain such that the position vector $\mathfrak{R}$:

$$\{ x_i \}_{\mathfrak{R},0} = \{ x_i (\mathrm{t} = 0)\} \}_{\mathfrak{R}}$$

is assigned to a fluid particle. Each of the moving fluid particles, resulting from the subdivision, marked and moving for $-\infty < t < +\infty$ is defined as a fluid element, that keeps its identity $0 \le t < \infty$ forever.

When kinematic considerations are carried out, only the motions of the in-dividual fluid elements are of interest. These considerations result for each fluid element in a separate and for the marked element characteristic path line. The computation of these path lines will be explained in the following. Here it is assumed that the flow field determining the fluid element motions is known.

As the velocity of a fluid element is only time-dependent, it follows from $\mathrm{d}\{ x_i \}_{\mathfrak{R}}/\mathrm{d}t = \{U_i\}_{\mathfrak{R}}$, that as a path line of a fluid element the frequency locus:

$$\{x_i\,(\mathrm{t})\}_{\Re} = \{x_i\}_{\Re,0} + \int_0^t \{U_i(t')\}_{\Re}\,\mathrm{d}t'$$

is to be understood. The position vector defined in this way for each moment in time t contains as a parameter the position vector of the particle defined at the time $t = 0$ *i.e.* $\Re$, *i.e.* $\{ x_1 \}_{\Re,0}$. Now the identity $\{U_i\} = \{U_i\}$ can be introduced into the considerations, *i.e.* at a certain moment in time t holds:

$$\frac{\mathrm{d}\{x_i\}_{\Re}}{\mathrm{d}t} = \{U_i\}_{\Re} = \{U_i\}$$

The equals sign between the substantial velocity $\{U_i\}_{\Re}$ and $\{U_i\}$ existing at the moment in time t indicates that the identity $\{x_i\}_{\Re} = \{x_i\}$ which exists at time t justifies equating the substantial velocity $\{U_i\}_{\Re}$ with the field size $\{U_i\}$. For the components $\{x_i\}_{\Re}$ of the particle motion it holds therefore:

$$\frac{\mathrm{d}\{x_i\}_{\Re}}{\mathrm{d}t} = \{U_i\} \text{ or } \frac{\mathrm{d}\{x_i\}_{\Re}}{\mathrm{d}t} = U_i$$

These differential equations have to be solved for $i = 1, 2, 3$ in order to determine the path lines of fluid elements. The differential quotient in the relation states that as a solution of the above differential equation the path line of a fluid element is obtained whose position was defined at the moment in time $t = 0$ with $\{x_i\}_{\Re,\,0}$.

The general way of proceeding when defining path lines will be explained and made clear by the example stated below. The components of the flow velocity field shall be given:

$$U_1 = x_1\,(1 + \mathrm{t}),\ U_2 = -x_2 \text{ und } U_3 = -x_3\mathrm{t}$$

When one inserts these statements on the velocity field in the above indicated differential equations for the path lines of a fluid element, one obtains:

$$\frac{\mathrm{d}\{x_1\}_{\Re}}{\mathrm{d}t} = x_1(1+t),$$

$$\frac{\mathrm{d}\{x_2\}_{\Re}}{\mathrm{d}t} = -x_2,$$

and

$$\frac{\mathrm{d}\{x_3\}_{\Re}}{\mathrm{d}t} = -x_3 t.$$

This law of differential equations can be solved now and results in the following solutions holding for the path lines of all fluid elements:

$$(x_1(t))_{\Re} = C_1 \exp\left[t + \frac{t^2}{2}\right]$$

$$(x_2(t))_{\Re} = C_2 \exp[-t]$$

$$(x_3(t))_{\Re} = C_3 \exp\left[-\frac{t^2}{2}\right].$$

When one considers a fluid element of interest which had the position coordinates (1, 1, 1)) at the time $t = 0$, then from the initial conditions for each of the equations in the introduced constants C_α result uniformly as:

$$C_1 = C_2 = C_3 = 1,$$

i.e. for the case considered here all integration constants are equal. For the path lines of this fluid element yields:

$$x_1\,(t)_{\Re} = \exp\left[t+\frac{t^2}{2}\right],$$

$$x_2\,(t)_{\Re} = \exp[-t]$$

$$x_3\,(t) = \exp\left[-\frac{t^2}{2}\right].$$

This path line is presented spatially in Fig. 4.2

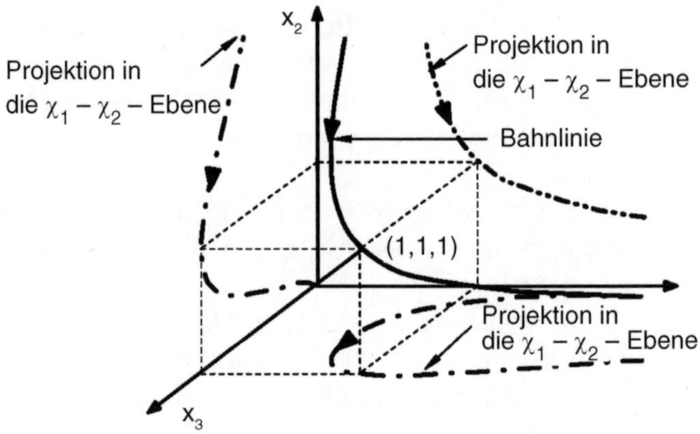

Fig. Spatial path line of the considered fluid element

When one selects a particle whose position at time $t = 0$ showed different position coordinates, the integration constants C_1 , C_2 , C_3 change accordingly and a different path line results. Thus the path line is an "'individual'" property of a fluid element which in the resulting equations is determined by the flow field and the position of the fluid element at the time $t = 0$.

The general solution for the position coordinates (x_i) ,0, which a fluid element takes at the time $t = 0$ is obtained when one inserts these coordinates in the general solutions for the path line coordinates for the determination of the integration constants $C_\alpha(\alpha = 1, 2, 3)$ This results in:

$$C_\alpha = (x_i),\, 0$$

and thus in the general solutions for the path line coordinates:

$$(x_1\,(t)) = (x_{\Re,0})_1 \exp\left[t+\frac{t^2}{2}\right]$$

$$(x_2\,(t)) = (x_{\mathfrak{R},0})_2 \exp[-t]$$

$$(x_3\,(t)) = (x_{\mathfrak{R},0})_1 \exp\left[-\frac{t^2}{2}\right].$$

These yield the space curves, with the time t as parameter, which represent each curve points of the path lines of fluid elements.

For further explanation, the following two-dimensional velocity field is also considered:[1]

$$U_1 = x_1\,,\ U_2 = x_2\,(1 + 2t) \text{ und } U_3 = 0.$$

With these data the following law of differential equations for the coordinates of the path lines of fluid elements can be formulated:

$$\frac{d(x_1)_{\mathfrak{R}}}{dt} = x_1, \frac{d(x_2)_{\mathfrak{R}}}{dt} = x_2(1+2t), \frac{d(x_3)_{\mathfrak{R}}}{dt} = 0$$

The solution of the third differential equation results in a constant that states in which plane the two-dimensional flow considerations are carried out. For the path line coordinates x_1 (t) and x_2 (t) it is computed:

$$(x_1)_{\mathfrak{R}} = C_1 \exp[t],\ (x_2)_{\mathfrak{R}} = C_2 \exp[t + t^2],\ (x_3)_{\mathfrak{R}} = C_3$$

When one computes the path line of the fluid element which at the time $t = 0$ took the coordinates (1,1,0), the result is:

$$(x_1)_{\mathfrak{R}} = \exp[t],\ (x_2)_{\mathfrak{R}} = \exp[t + 1],\ (x_3)_{\mathfrak{R}} = 0$$

When one resolves the equation obtained for (x_1) with respect to time, it yields:

$$t = \ln(x_1)_{\mathfrak{R}}$$

When inserted in the solution for $(x_2\,)_{\mathfrak{R}}$ for two-dimensional path lines in the plane $x_1 - x_2$ – the following functional relation between $(x_1)_{\mathfrak{R}}$ and $(x_2)_{\mathfrak{R}}$ yields:

$$(x_2)_{\mathfrak{R}} = (x_1)_{\mathfrak{R}}^{(1+\ln(x_1)_{\mathfrak{R}})}$$

Sweeping paths of locally injected tracer materials

As mentioned before, it is usual in experimental fluid mechanics to gain qualitative insight in a flow process by injecting a continuous fluid tracer at a fixed position. This leads to a marked "fluid thread" which is carried with the flow and thus marks/traces the course of the flow. When the exact course of the flow is of interest, quantitative evaluations of the location coordinates of sweeping paths of locally installed tracer materials are required. These evaluations can, based on the derivations stated below, be carried out with methods of flow kinematics.

Attention is drawn to the fact that this flow field is not source-free , thus it violates the requirements/demands of the continuity equation. This is, however, insignificant for the purely kinematic considerations mentioned here.

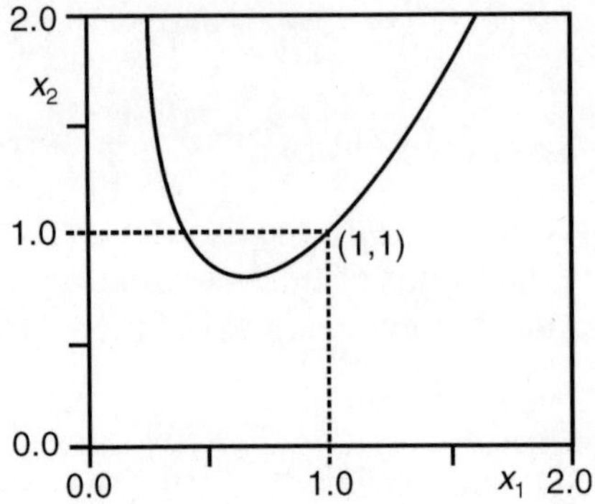

Fig. Path line of the flow running parallel to the plane $x_1 - x_2$

A fluid particle marked with a tracer, *e.g.* an air particle or any other gas particle marked with smoke, or a water or fluid particle marked with colour, which at the time t is located at the position $\{x_i\} = \{x_i(t)\}_{\Re}$ must have passed the injection point for the tracer at a moment in time $(t - \tau)$, in order to be present as a marked particle at the point $\{x_i\}$ *i.e.* it holds :

$$\{x_i(t)\}_{\Re} = \{x_i(t-\tau)\}_S$$

Hence the way covered by a marked fluid element up to the time t can be computed as path line of the element that fulfills the condition *i.e.* a path line with the initial condition that for $t = \tau$ the fluid element held the position of the location coordinates of the injection point. The sweeping path thus is composed of the sum of the path lines of individual particles. For each individual marked particle of a sweeping path a parameter τ is introduced, which for $0 \le \tau \le t$ covers all parts of a sweeping path. It is therefore important to vary the parameter τ in the solution equations in order obtain the entire sweeping path.

The above short explanations shall be made clear again by way of an example, which is handled on the basis of the three-dimensional velocity field used above:

$$U_1 = x_1 (1 + t),\ U_2 = -x_2 \text{ and } U_3 = -x_3 t$$

This velocity field yields the law of differential equations for the motion of a fluid element in space :

$$\frac{d(x_1)_S}{dt} = x_1(1+t),\ \frac{d(x_2)_S}{dt} = x_1,\ \frac{d(x_3)_S}{dt} = -x_3 t.$$

As a solution one obtains for the components$(x_1)_S$, $(x_2)_S$ and $(x_3)_S$ according to equation:

[1]The index s signifies that the location coordinate of the sweeping path is meant.

$$(x_1)_S = C_1 \exp\left[t(1+\frac{t}{2})\right],\ (x_2)_S = C_2 \exp[-t],\ (x_3)_S = C_3 \exp\left[-\frac{t^2}{2}\right].$$

When one inserts now the initial conditions, that $(x_1)_S = (x_1)_{t=\tau} = 1$, $(x_2)_S = (x_2)_{t=\tau} = 1$, $(x_3)_S = (x_3)_{t=\tau} = 1$ was present for $t = \tau$, *i.e.* that the position (1, 1, 1) serves as an injection point of the tracer, one obtains:

$$C_1 = \exp\left[-\tau(1+\frac{\tau}{2})\right];$$

$$C_2 = \exp\,[\tau] \text{ and } C_3 = \exp\left[\frac{\tau^2}{2}\right].$$

Inserted in the solutions for $(x_1)_S$, $(x_2)_S$ and $(x_3)_S$ the equation of the frequency locus defined as sweeping path yields for all times:[2] :

$$(x_1)_S = \exp\left[t(1+\frac{t}{2}) - \tau(1+\frac{\tau}{2})\right],$$

$$(x_1)_S = \exp\,[-(t-\tau)],$$

$$(x_1)_S = \exp\left[1\frac{1}{2}(t^2-\tau^2)\right]$$

When one wants to make visible the course of a sweeping path at a moment in time t (partly), one has to insert the value of t in the above equation in order to obtain in this way the equation of a space curve, with τ as a parameter. Here t is determined by the period of time $[\tau_1, \tau_2]$ of the tracer injection in (1, 1, 1) with $-\infty < \tau_1 < \tau_2 < t$. For $\tau_1 \to -\infty$, $\tau_2 = t$ and $t = 0$ yields:

$$(x_1)_S = \exp\left[-\tau(1+\frac{\tau}{2})\right],$$

$$(x_2)_S = \exp\,[\tau]\,, -\infty < \tau < 0$$

$$(x_3)_S = \exp\left[\frac{\tau^2}{2}\right]$$

The course of this space curve is shown in 4.4. It indicates the sweeping path existing at the moment in time $\tau = 0$ (made visible from $\tau = -\infty$ to $\tau = 0$, the projections of the sweeping path into the main level of the Cartesian coordinate system are also introduced.

When one compares the equation for the sweeping path fixed by the space point (1, 1, 1) with the equations for the path line of a fluid element, stated for the same flow field, one realizes that path lines and sweeping paths are not identical for non-stationary flows. Only in the case of a stationary flow field path lines and sweeping paths are identical, as can be shown easily by the following considerations.

As a space curve is concerned here, the statement in x_1, x_2, x_3 – coordinates is appropriate. The definition x_S indicates that the location coordinates of a sweeping path are meant.

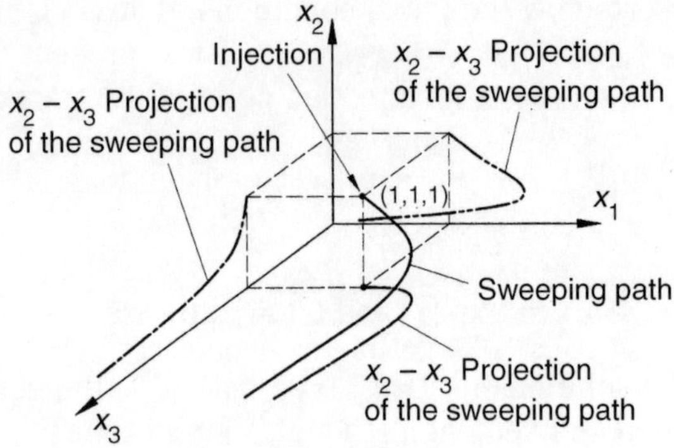

Fig. Sweeping path for the moment in time t = 0. with fluid tracer injec-tions between $\tau = -\infty$ and $\tau = 0$ at the position (1, 1, 1)

Considering the stationary velocity field:

$$U_1 = 2x_1\,,\; U_2 = -x_2,\; U_3 = -x_3$$

one obtains for the path line of a fluid element the following differential equation:

$$\frac{d(x_1)_{\Re}}{dt} = 2x_1, \frac{d(x_2)_{\Re}}{dt} = -x_2, \frac{d(x_3)_{\Re}}{dt} = x_3$$

for $t = 0$ it shall be assumed that $(x_1)_{\Re} = (x_2)_{\Re} = (x_3)_{\Re} = 1$ so that in the solution

$$(x_1) = C_1 \exp[2t]\,,\; (x_2)_{\Re} = C_2 \exp[-t]\,,\; (x_3) = C_3 \exp[-t]$$

holds and thus the path line is stated as follows:

$$(x_1)_{\Re} = \exp[2t]\,,\; (x_2)_{\Re} = \exp[-t],\; (x_3)_{\Re} = \exp[-t]$$

with $-\infty < t < \infty$. For the computation of the sweeping paths the solution can be employed again and C_1, C_2, C_3 can be computed such that it is claimed that at the time $t = \tau$ holds:

$$(x_1(t = \tau))_S = 1,\; (x_2(t = \tau))_S = 1,\; (x_3(t = \tau))_S = 1$$

Therefore it holds:

$$C_1 = \exp[-2\tau]\,,\; C_2 = \exp[\tau]\,,\; C_3 = \exp[\tau]$$

or as an equation for the sweeping path:

$$(x_1)_S = \exp[2(t - \tau)],\; (x_2)_S = \exp[-(t - \tau)],\; (x_3)_S = \exp[-(t - \tau)]$$

thus t being defined, and the range of values of t is defined by the period of time of the tracer injection. In the case that tracer substance is injected at all times, *i.e.* $-\infty < \tau < \infty$, equation yield the same curve. When the tracer injection is limited in time, one obtains as a visible sweeping path a corresponding part of the path line.

$$U_1 = x_1,\; U_2 = x_2(1 + 2t) \text{ and } U_3 = 0$$

which leads to the differential equations:

$$\frac{d(x_1)_{\Re}}{dt} = x_{\Re}, \frac{d(x_2)_{\Re}}{dt} = x_{\Re}(1+2t), \frac{d(x_3)_{\Re}}{dt} = 0$$

with the solutions:

$$(x_1)_{\Re} = C_1 \exp[t],\ (x_2)_{\Re} = C_2 \exp[t(t+1)],\ (x_3)_{\Re} = C_3$$

On the other hand, when one demands that the particle located at the time t in the point x_1, x_2, x_3 passes the injection point (1, 1, 0) of a tracer at the moment in time τ the integration constants C_1, C_2 and C_3 can be obtained from the following conditional equations:

$$C_1 \exp[\tau] = 1,\ C_2 \exp[\tau(\tau+1)] = 1,\ C_3 = 0$$

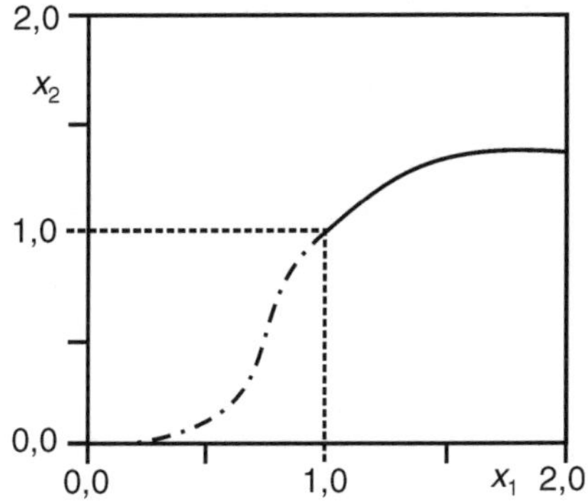

Fig. Sweeping path in the plane $x_1 - x_2$ (fully drawn line corresponds to $-\infty \le \tau \le 0$, broken line to $0 \le \tau \le \infty$)

These "'constants"' can now be inserted in equation again, where $C_3 = 0$ signifies that the sweeping path runs in the plane $x_1 - x_2$, *i.e.* in the plane $x_3 = 0$, and is described there by the following equations for the position coordinates $(x_1)_S$ and $(x_2)_S$

$$(x_1)_S = \exp[t - \tau],$$
$$(x_2)_S = \exp[t(t+1) - \tau(\tau+1)]$$

For the moment in time $t = 0$ the course of the sweeping path is:

$$(x_1)_S = \exp[-\tau],$$
$$(x_2)_S = \exp[-\tau(\tau+1)]$$

in the domain $-\infty < \tau < \infty$.

From the equation for $(x_1)_S$ follows:

$$t = -\ln(x_1)_S$$

Inserted in the equation for $(x_2)_s$ at the time $t = 0$ the existing course of the sweeping path made visible by the injection of tracer material in $-\infty < \tau < \infty$ results in the plane $x_1 - x_2$

$$(x_2)_S = (x_1)_S^{(1-1n(x_1)_S)} \text{ in the domain } 0 < (x1)_S < \infty$$

$$(x_2)_S = [(x_1)_S]^{(1-1n(x_1)_S)}$$

VORTEX MOTION

Vortex (vôr'teks) [key], mass of fluid in whirling or rotary motion. To simplify the analysis, vortex motion usually describes motions in a frictionless fluid. In such cases the absence of friction would make it impossible to create or to destroy vortex motion. Motion in such a fluid would be a permanent flow pattern; the velocity of the fluid element instantaneously passing through a given point in space would be constant in time. Lines drawn so that their direction is that of the axis of rotation of the fluid are called vortex lines, and if these lines close on themselves they are called vortex rings. Hermann von Helmholtz was probably the first to investigate the properties of vortex motion; Lord Kelvin developed a theory of the material atom as a vortex ring; and J. C. Maxwell worked out a theory of electromagnetism, assuming that every magnetic tube of force was a vortex with an axis of rotation coinciding with the direction of the force. Many properties have been mathematically proved for the perfect frictionless fluid. In practice, however, their full realization is impossible because no frictionless fluid exists. To maintain a vortex motion a continuous energy supply to overcome friction is needed. A smoke ring is a familiar example of a typical vortex motion in which the medium is air. In this case the rings are stable for a short time because of the comparatively slight friction in air. An illustration of vortex motion in a liquid medium is the small whirlpool formed by water as it drains from a wash basin. In nature, illustrations of vortical motion on a larger scale are seen in waterspouts, whirlpools, and tornadoes. Investigations of sunspots reveal enormous vortices in the gases surrounding them. The principles of vortex motion are applied in aerodynamics, *e.g.*, to explain the movement of air behind the trailing edge of a wing.

FLUID DYNAMICS

Fluid dynamics is the sub-discipline of fluid mechanics dealing with fluids (liquids and gases) in motion. It has several subdisciplines itself, including aerodynamics (the study of gases in motion) and hydrodynamics (the study of liquids in motion). Fluid dynamics has a wide range of applications, including calculating forces and moments on aircraft, determining the mass flow rate of petroleum through pipelines, predicting weather patterns, understanding nebulae in interstellar space and reportedly modelling fission weapon detonation. Some of its principles are even used in traffic engineering, where traffic is treated as a continuous fluid.

Fluid dynamics offers a systematic structure that underlies these practical disciplines and that embraces empirical and semi-empirical laws, derived from flow measurement, used to solve practical problems. The solution of a fluid dynamics problem typically involves calculation of various properties of the fluid, such as velocity, pressure, density, and temperature, as functions of space and time.

EQUATIONS OF FLUID DYNAMICS AND AERODYNAMICS

The foundational axioms of fluid dynamics are the conservation laws, specifically, conservation of mass, conservation of momentum (also known as

Newton's second law and third law), and conservation of energy. These are based on classical mechanics and are modified in quantum mechanics and general relativity. They are expressed using the Reynolds transport theorem.

In addition to the above, fluids are assumed to obey the continuum assumption. Fluids are composed of molecules that collide with one another and solid objects. However, the continuum assumption considers fluids to be continuous, rather than discrete. Consequently, properties such as density, pressure, temperature, and velocity are taken to be well-defined at infinitely small points, and are assumed to vary continuously from one point to another. The fact that the fluid is made up of discrete molecules is ignored.

For fluids which are sufficiently dense to be a continuum, do not contain ionized species, and have velocities small in relation to the speed of light, the momentum equations for Newtonian fluids are the Navier-Stokes equations, which are non-linear differential equations that describe the flow of a fluid whose stress depends linearly on velocity and on pressure. The unsimplified equations do not have a general closed-form solution, so they are only of use in computational fluid dynamics or when they can be simplified. The equations can be simplified in a number of ways, all of which make them easier to solve. Some of them allow appropriate fluid dynamics problems to be solved in closed form.

In addition to the mass, momentum and energy conservation equations, a thermodynamical equation of state giving the pressure as a function of other thermodynamic variables for the fluid is required to completely specify the problem.

An example of this would be the perfect gas equation of state:

$$p = \frac{\rho R_u T}{M}$$

where p is pressure, ρ is density, R_u is the gas constant, M is the molecular mass and T is temperature.

COMPRESSIBLE VS INCOMPRESSIBLE FLOW

All fluids are compressible to some extent, that is changes in pressure or temperature will result in changes in density. However, in many situations the changes in pressure and temperature are sufficiently small that the changes in density are negligible. In this case the flow can be modelled as an incompressible flow. Otherwise the more general compressible flow equations must be used.

Mathematically, incompressibility is expressed by saying that the density ρ of a fluid parcel does not change as it moves in the flow, *i.e.*

$$\frac{D\rho}{Dt} = 0,$$

where D/Dt is the convective derivative. This additional constraint simplifies the governing equations, especially in the case when the fluid has a uniform density.

For flow of gases, to determine whether to use compressible or incompressible fluid dynamics, the Mach number of the problem is evaluated. As a rough guide, compressible effects can be ignored at Mach numbers below approximately 0.3. For liquids, whether the incompressible assumption is valid depends on the fluid properties (specifically the critical pressure and temperature of the fluid) and the flow conditions (how close to the critical pressure the actual flow pressure becomes). Acoustic problems always require allowing compressibility, since sound waves are compression waves involving changes in pressure and density of the medium through which they propagate.

VISCOUS VS INVISCID FLOW

Viscous problems are those in which fluid friction has significant effects on the solution.

The Reynolds number can be used to evaluate whether viscous or inviscid equations are appropriate to the problem.

Stokes flow is flow at very low Reynolds numbers, such that inertial forces can be neglected compared to viscous forces.

On the contrary, high Reynolds numbers indicate that the inertial forces are more significant than the viscous (friction) forces. Therefore, we may assume the flow to be an inviscid flow, an approximation in which we neglect viscosity at all, compared to inertial terms.

This idea can work fairly well when the Reynolds number is high, even if certain problems, such as those involving boundaries, may require that viscosity be included. Viscosity often cannot be neglected near boundaries because the no-slip condition can generate a region of large strain rate (a Boundary layer) which enhances the effect of even a small amount of viscosity, generating vorticity. Therefore, to calculate net forces on bodies (such as wings) we should use viscous equations. As illustrated by d'Alembert's paradox, a body in an inviscid fluid will experience no force. The standard equations of inviscid flow are the Euler equations. Another often used model, especially in computational fluid dynamics, is to use the Euler equations far from the body and the boundary layer equations, which incorporate viscosity, close to the body.

The Euler equations can be integrated along a streamline to get Bernoulli's equation. When the flow is everywhere irrotational and inviscid, Bernoulli's equation can be used throughout the field.

STEADY VS UNSTEADY FLOW

When all time derivatives of a flow field vanish, the flow is considered steady. Otherwise, it is called unsteady. Whether a particular flow is steady or unsteady, can depend on the chosen frame of reference. For instance, laminar flow over a sphere is steady in the frame of reference that is stationary with respect to the sphere. In a frame of reference that is stationary with respect to the background flow, the flow is unsteady.

Turbulent flows are unsteady by definition. A turbulent flow can, however, be statistically stationary. According to Pope, Turbulent Flows, Cambridge university press, page 75:

The random field U(x,t) is statistically stationary if all statistics are invariant under a shift in time.

This roughly means that all statistical properties are constant in time. Often, the mean field is the object of interest, and this is constant too in a statistically stationary flow.

Steady flows are often more tractable than otherwise similar unsteady flows. The governing equations of a steady problem have one dimension less (time) than the governing equations of the same problem without taking advantage of the steadiness of the flow field.

Although strictly unsteady flows, time-periodic problems can often be solved by the same techniques as steady flows. For this reason, they can be considered to be somewhere between steady and unsteady.

LAMINAR VS TURBULENT FLOW

Turbulence is flow dominated by recirculation, eddies, and apparent randomness. Flow in which turbulence is not exhibited is called laminar. It should be noted, however, that the presence of eddies or recirculation does not necessarily indicate turbulent flow—these phenomena may be present in laminar flow as well. Mathematically, turbulent flow is often represented via Reynolds decomposition, in which the flow is broken down into the sum of a steady component and a perturbation component.

It is believed that turbulent flows obey the Navier-Stokes equations. Direct numerical simulation (DNS), based on the incompressible Navier-Stokes equations, makes it possible to simulate turbulent flows with moderate Reynolds numbers (restrictions depend on the power of computer and efficiency of solution algorithm). The results of DNS agree with the experimental data.

Most flows of interest have Reynolds numbers too high for DNS to be a viable option (see: Pope), given the state of computational power for the next few decades. Any flight vehicle large enough to carry a human (L > 3 m), moving faster than 72 km/h (20 m/s) is well beyond the limit of DNS simulation (Re = 4 million). Transport aircraft wings (such as on an Airbus A300 or Boeing 747) have Reynolds numbers of 40 million (based on the wing chord). In order to solve these real life flow problems, turbulence models will be a necessity for the foreseeable future. Reynolds-averaged Navier-Stokes equations combined with turbulence modeling provides a model of the effects of the turbulent flow, mainly the additional momentum transfer provided by the Reynolds stresses, although the turbulence also enhances the heat and mass transfer. Large eddy simulation also holds promise as a simulation methodology, especially in the guise of detached eddy simulation (DES), which is a combination of turbulence modeling and large eddy simulation.

NEWTONIAN VS NON-NEWTONIAN FLUIDS

Sir Isaac Newton showed how stress and the rate of change of strain are very close to linearly related for many familiar fluids, such as water and air. These Newtonian fluids are modeled by a coefficient called viscosity, which depends on the specific fluid.

However, some of the other materials, such as emulsions and slurries and some visco-elastic materials (eg. blood, some polymers), have more complicated *non-Newtonian* stress-strain behaviours. These materials include *sticky liquids* such as latex, honey, and lubricants which are studied in the sub-discipline of rheology.

OTHER APPROXIMATIONS

There are a large number of other possible approximations to fluid dynamic problems. Some of the more commonly used are listed below.

- The Boussinesq approximation neglects variations in density except to calculate buoyancy forces. It is often used in free convection problems where density changes are small.
- Lubrication theory exploits the large aspect ratio of the domain to show that certain terms in the equations are small and so can be neglected.
- Slender-body theory is a methodology used in Stokes flow problems to estimate the force on, or flow field around, a long slender object in a viscous fluid.
- The shallow-water equations can be used to describe a layer of relatively inviscid fluid with a free surface, in which surface gradients are small.
- Darcy's law is use for flow in porous media, and works with variables averaged over several pore-widths.
- In rotating systems, the quasi-geostrophic approximation assumes an almost perfect balance between pressure gradients and the Coriolis force. It is useful in the study of atmospheric dynamics.

EULER'S FORMULA

Euler's formula, named after Leonhard Euler, is a mathematical formula in complex analysis that shows a deep relationship between the trigonometric functions and the complex exponential function. (Euler's identity is a special case of the Euler formula.)

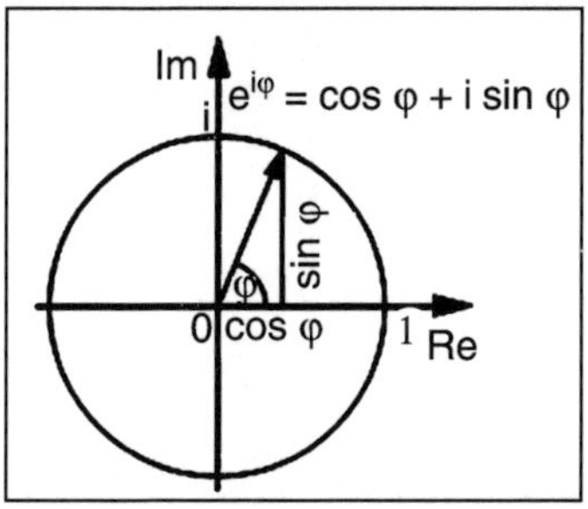

Euler's formula states that, for any real number *x*,

$$e^{ix} = \cos(x) + i \sin(x)$$

where,

e is the base of the natural logarithm

i is the imaginary unit

cos and sin are trigonometric functions.

Richard Feynman called Euler's formula "our jewel" and "the most remarkable formula in mathematics".

HISTORY

Euler's formula was proven for the first time by Roger Cotes in 1714 in the form

$$\text{Ln}\ (\cos (x) + 0 \sin (x)) = ix$$

(where "ln" means natural logarithm, *i.e.* log with base *e*).

It was Euler who published the equation in its current form in 1748, basing his proof on the infinite series of both sides being equal. Neither of these men saw the geometrical interpretation of the formula: the view of complex numbers as points in the complex plane arose only some 50 years later (see Caspar Wessel). Euler considered it natural to introduce students to complex numbers much earlier than we do today. In his elementary algebra text book, *Elements of Algebra*, he introduces these numbers almost at once and then uses them in a natural way throughout.

APPLICATIONS IN COMPLEX NUMBER THEORY

This formula can be interpreted as saying that the function e^{ix} traces out the unit circle in the complex number plane as x ranges through the real numbers. Here, x is the angle that a line connecting the origin with a point on the unit circle makes with the positive real axis, measured counter clockwise and in radians. The formula is valid only if sin and cos take their arguments in radians rather than in degrees.

The original proof is based on the Taylor series expansions of the exponential function e^z (where z is a complex number) and of sin x and cos x for real numbers x (see below). In fact, the same proof shows that Euler's formula is even valid for all complex numbers z.

Euler's formula can be used to represent complex numbers in polar coordinates. Any complex number z = x + iy can be written as,

$$z = x + iy = |z|(\cos\phi + i\sin\phi) = |z|e^{i\phi}$$
$$z = x - iy = |z|(\cos\phi - i\sin\phi) = |z|e^{-i\phi}$$

where,

x = Re {z} the real part

y = Im {z}the imaginary part

$|z|\sqrt{x^2 + y^2}$ the magnitude of *z*

and ϕ is the *argument* of *z*— the angle between the *x* axis and the vector *z* measured counterclockwise and in radians — which is defined up to addition of 2P.

Now, taking this derived formula, we can use Euler's formula to define the logarithm of a complex number. To do this, we also use the facts that,

$$a = e^{\ln(a)}$$

and,

$$e^a\, e^b = e^{a+b}$$

both valid for any complex numbers *a* and *b*.

Therefore, one can write:

$$z = |z|e^{i\phi} = e^{\ln|z|}ee^{i\phi} = ee^{\ln|z|+i\phi}$$

for any $z \neq 0$.

Taking the logarithm of both sides shows that:

$$\ln z = \ln|z| + i\phi.$$

and in fact this can be used as the definition for the complex logarithm. The logarithm of a complex number is thus a multi-valued function, due to the fact that ϕ is multi-valued.

Finally, the other exponential law,

$$(e^a)^k = e^{ak},$$

which can be seen to hold for all integers *k*, together with Euler's formula, implies several trigonometric identities as well as de Moivre's formula.

RELATIONSHIP TO TRIGONOMETRY

Euler's formula provides a powerful connection between analysis and trigonometry, and provides an interpretation of the sine and cosine functions as weighted sums of the exponential function:

$$\cos x = \mathrm{Re}\left[e^{ix}\right] = \frac{e^{ix} + e^{-ix}}{2}$$

$$\sin x = \mathrm{Im}\left[e^{ix}\right] = \frac{e^{ix} - e^{-ix}}{2i}$$

The two equations above can be derived by adding or subtracting Euler's formulas:

$$e^{ix} = \cos x + i \sin x$$

$$e^{-ix} = \cos(-x) + i \sin(-x) = \cos x - i \sin x$$

and solving for either cosine or sine.

These formulas can even serve as the definition of the trigonometric functions for complex arguments x. For example, letting $x = iy$, we have:

$$\cos(iy) = \frac{e^{-y} + e^{y}}{2} = \cosh(y)$$

$$\sin(iy) = \frac{e^{-y} - e^{y}}{2i} i \sinh(y).$$

Complex exponentials can simplify trigonometry, because they are easier to manipulate than their sinusoidal components. One technique is simply to convert sinusoids into equivalent expressions in terms of exponentials. After the manipulations, the simplified result is still real-valued.

For example:

$$\begin{aligned}\cos(x)\cdot\cos(y) &= \frac{\left(e^{ix}+e^{-ix}\right)}{2}\cdot\frac{\left(e^{iy}+e^{-iy}\right)}{2} \\ &= \frac{e^{i(x+y)}+e^{i(x-y)}+e^{i(-x+y)}+e^{i(-x-y)}}{4} \\ &= \frac{e^{i(x+y)}+e^{i(-x-y)}}{4}+\frac{e^{i(x-y)}+e^{i(-x+y)}}{4} \\ &= \frac{\cos(x+y)}{2}+\frac{\cos(x-y)}{2}.\end{aligned}$$

Another technique is to represent the sinusoids in terms of the real part of a more complex expression, and perform the manipulations on the complex expression. For example:

$$\begin{aligned}\cos(x\cdot n)+\cos(x(n-2)) &= \mathrm{Re}\left\{e^{ixn}+e^{ix(n-2)}\right\} \\ &= \mathrm{Re}\left\{e^{ix(n-1)}\cdot+\left(e^{ix}+e^{-ix}\right)\right\} \\ &= \mathrm{Re}\left\{e^{ix(n-1)}\cdot 2\cos(x)\right\} \\ &= \cos(x(n-1))\cdot 2\cos(x).\end{aligned}$$

OTHER APPLICATIONS

In differential equations, the function e^{ix} is often used to simplify derivations, even if the final answer is a real function involving sine and cosine. Euler's identity is an easy consequence of Euler's formula.

In electrical engineering and other fields, signals that vary periodically over time are often described as a combination of sine and cosine functions (see Fourier analysis), and these are more conveniently expressed as the real part of exponential functions with imaginary exponents, using Euler's formula. Also, phasor analysis of circuits can include Euler's formula to represent the impedance of a capacitor or an inductor.

PROOFS

Using Taylor series

Here is a proof of Euler's formula using Taylor series expansions as well as basic facts about the powers of *i*:

$i^0 = 1, \quad i^1 = i, \quad i^2 = -1, \quad i^3 = -i,$

$i^4 = 1, \quad i^5 = i, \quad i^6 = -1, \quad i^7 = -i,$

and so on. The functions e^x, $\cos(x)$ and $\sin(x)$ (assuming x is real) can be expressed using their Taylor expansions around zero:

$$e^x = 1 + x + \frac{x^2}{2!} + \frac{x^3}{3!} + \ldots$$

$$\cos x = 1 - \frac{x^2}{2!} + \frac{x^4}{4!} - \frac{x^6}{6!} + \ldots$$

$$\sin x = x - \frac{x^3}{3!} + \frac{x^5}{5!} - \frac{x^7}{7!} + \ldots$$

For complex z we *define* each of these function by the above series, replacing x with z. This is possible because the radius of convergence of each series is infinite. We then find that,

$$\begin{aligned} e^{iz} &= 1 + iz + \frac{(iz)^2}{2!} + \frac{(iz)^3}{3!} + \frac{(iz)^4}{4!}\frac{(iz)^5}{5!} + \frac{(iz)^6}{6!} + \frac{(iz)^7}{7!} + \frac{(iz)^8}{8!} + \ldots \\ &= 1 + iz - \frac{z^2}{2!} + \frac{iz^3}{3!} + \frac{z^4}{4!} + \frac{iz^5}{5!} + \frac{z^6}{6!} + \frac{iz^7}{7!} + \frac{z^8}{8!} + \ldots \\ &= \left(1 - \frac{z^2}{2!} + \frac{z^4}{4!} - \frac{z^6}{6!} + \frac{z^8}{8!} - \ldots\right) + i\left(z - \frac{z^3}{3!} + \frac{z^5}{5!} - \frac{z^7}{7!} + \ldots\right) \end{aligned}$$

The rearrangement of terms is justified because each series is absolutely convergent. Taking $z = x$ to be a real number gives the original identity as Euler discovered it.

USING CALCULUS

Define the function f by,

$$f(x) = \frac{\cos x + i \sin x}{e^{ix}}.$$

This is allowed since the equation,

$$e^{ix} \cdot e^{-ix} = e^{0} = 1$$

implies that e^{ix} is never zero.

The derivative of f, according to the quotient rule, is:

$$f'(x) = \frac{(-\sin x + i\cos x)\cdot e^{ix} - (\cos x + i\sin x)\cdot i\cdot e^{ix}}{\left(e^{ix}\right)^{2}}$$

$$= \frac{-\sin x\cdot e^{ix} - i^{2}\sin x\cdot e^{ix}}{\left(e^{ix}\right)^{2}}$$

$$= \frac{\left(-1 - i^{2}\right)\cdot \sin x\cdot e^{ix}}{\left(e^{ix}\right)^{2}}$$

$$= \frac{\left(-1 - (-1)\right)\cdot \sin x\cdot e^{ix}}{\left(e^{ix}\right)^{2}}$$

$$= 0.$$

Therefore, f must be a constant function. Thus,

$$\frac{\cos x + i\sin x}{e^{ix}} = f(x) = f(0) = \frac{\cos 0 + i\sin 0}{e^{0}} = 1.$$

Rearranging, it follows that,

$$\cos x + i \sin x = e^{ix}.$$

DYNAMICAL SYSTEM OF MOTION

It is often useful to represent the motion of objects pictorially on a position vs time graph. Such a graph takes the form of a line. Each point on the line specifies the position (relative to the origin) of an object at a given instant in time. The simplest "motion" is that of a stationary object, whose velocity and acceleration are zero. Since the position of such an object doesn't change its position vs time graph is a straight, horizontal line, as shown in blue in the Figure.

The next simplest motion, is that of an object going at constant velocity (hence zero acceleration). Its position vs time graph is again a straight line, but with non-zero slope. In fact the slope of the line (the rise over the run) equals the speed of the object. The red line in Fig shows the position time graph of an object moving at 2 m/s along the positive x-axis. Since it is a straight line, the slope can be calculated between any two points on the graph. If the object were moving with constant velocity along the negative x-axis, the line would be sloped in the opposite direction: the value of the x-coordinate would decrease as time increased. This is illustrated by the green line.

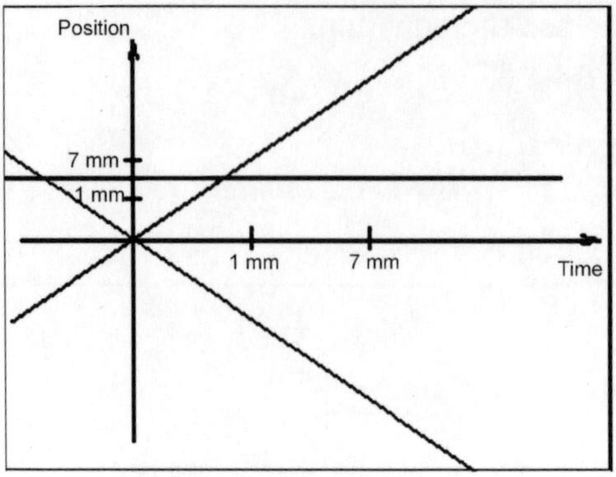

Fig. Position-Time Graph

FREE FALL MOTION

One important elementary example of motion is that of free fall, which is the motion of an object only under the influence of gravity. It is a remarkable fact, first discovered over 300 years ago by Galileo and others, that objects in free fall motion descend at the same rate, *independent of their mass*, as long as the effects of air resistance can be neglected. This means that under free fall all objects have the same constant acceleration, which in the metric system is 9.8 m/s^2 at sea level, directed towards the center of the earth.

However, if you now put the piece of paper on top of the text book, and drop them together, you will see quite dramatically that the paper and the book drop at precisely the same rate. In the second experiment, clearly the text book, which is less affected by air resistance, is shielding the piece of paper from the affects of air resistance so that both are essentially moving only under the influence of gravity. Consequently, as stated above, they accelerate at the same rate.

Graphs of how the position, velocity, and acceleration of objects under free fall motion change in time follow.

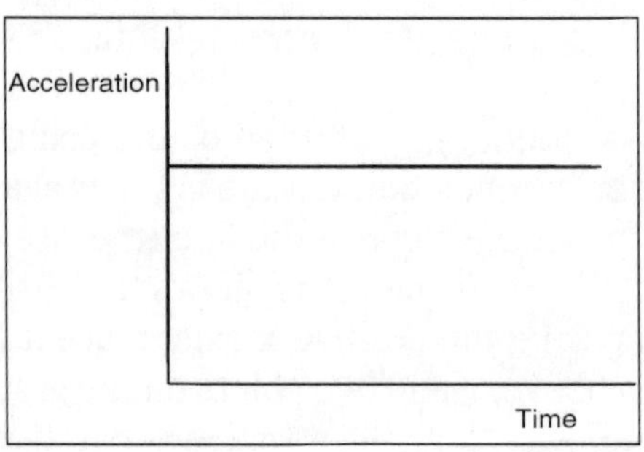

Fig. Free fall acceleration

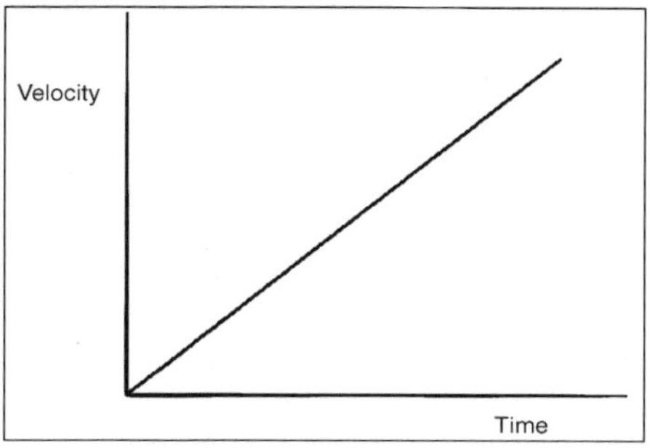

Fig. Free fall velocity

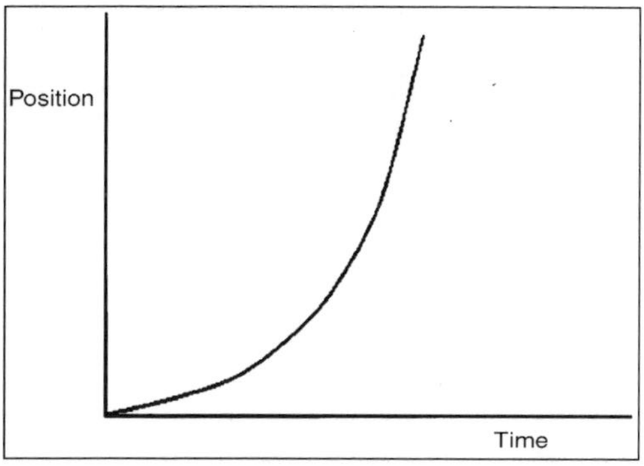

Fig. Free fall position

NEWTONIAN KINETIC ENERGY

KINETIC ENERGY OF RIGID BODIES

In classical mechanics, the kinetic energy of a "point object" (a body so small that its size can be ignored), or a non-rotating rigid body, is given by the equation $E_k = \frac{1}{2}mv^2$ where m is the mass and v is the speed of the body.

For example - one would calculate the kinetic energy of an 80 kg mass traveling at 18 meters per second (40 mph) as,

$$\frac{1}{2} \cdot 80 \cdot 18^2 = 12{,}960\,\text{joules}..$$

Note that the kinetic energy increases with the square of the speed. This means, for example, that if you are traveling twice as fast, you will have four times as much kinetic energy. As a result of this, a car traveling twice as fast requires four times as much distance to stop.

Thus, the kinetic energy can be calculated using the formula:

$$E_k = \frac{1}{2}mv^2$$

For the *translational kinetic energy* of a body with constant mass m, whose center of mass is moving in a straight line with speed v, as seen above is equal to

$$E_t = \frac{1}{2} mv^2$$

where:

m is mass of the body

v is speed of the center of mass of the body.

Thus kinetic energy is a relative measure and no object can be said to have a unique kinetic energy. A rocket engine could be seen to transfer its energy to the rocket ship or to the exhaust stream depending upon the chosen frame of reference. But the total energy of the system, *i.e.* kinetic energy, fuel chemical energy, heat energy etc, will be conserved regardless of the choice of measurement frame.

The kinetic energy of an object is related to its momentum by the equation:

$$E_k = \frac{p^2}{2m}$$

Derivation and Definition

The work done accelerating a particle during the infinitesimal time interval dt is given by the dot product of *force* and *displacement*:

$$F \cdot dx = F \cdot vdt = \frac{dp}{dt} \cdot vdt = v \cdot dp = v \cdot d(mv)$$

Applying the product rule we see that:

$$d(v \cdot v) = (dv) \cdot v + v \cdot (dv) = 2(v \cdot dv)$$

Therefore (assuming constant mass), the following can be seen:

$$v \cdot d(mv) = \frac{m}{2} d(v \cdot v) = \frac{m}{2} dv^2 = d\left(\frac{mv^2}{2}\right)$$

Since this is a total differential (that is, it only depends on the final state, not how the particle got there), we can integrate it and call the result kinetic energy:

$$E_k = \int F \cdot dx = \int v \cdot dp = \frac{mv^2}{2}$$

This equation states that the kinetic energy (E_k) is equal to the integral of the dot product of the velocity (v) of a body and the infinitesimal change of the body's momentum (p). It is assumed that the body starts with no kinetic energy when it is at rest (motionless).

KINETIC ENERGY OF SYSTEMS

For a single point, or a rigid body that is not rotating, the kinetic energy goes to zero when the body stops.

However, for systems containing multiple independently moving bodies, which may exert forces between themselves, and may (or may not) be rotating; this is no longer true.

This energy is called 'internal energy'.

The kinetic energy of the system at any instant in time is simply the sum of the kinetic energies of the masses- including the kinetic energy due to the rotations.

An example would be the solar system. In the center of mass frame of the solar system, the Sun is (almost) stationary, but the planets and planetoids are in motion about it. Thus even in a stationary center of mass frame, there is still kinetic energy present.

However, recalculating the energy from different frames would be tedious, but there is a trick. The kinetic energy of the system from a different inertial frame can be calculated simply from the sum of the kinetic energy in the center of mass frame and adding on the energy that the total mass of bodies in the center of mass frame would have if it were moving at the relative speed between the two frames.

This may be simply shown: let V be the relative speed of the frame *k* from the center of mass frame *i* :

$$E_k = \int \frac{v_k^2 dm}{2} = \int \frac{(v_i + V)^2 dm}{2} = \int \frac{(v_i^2 + 2v_i V + V^2) dm}{2}$$

$$= \int \frac{v_i^2 dm}{2} + v \int v_i dm + \frac{V^2}{2} \int dm$$

However, let

$$\int \frac{v_i^2 dm}{2} = E_i$$

the kinetic energy in the center of mass frame, $\int v_i dm$ would be simply the total momentum which is by definition zero in the center of mass frame, and let the total mass:

$$\int dm = M.$$

Substituting, we get:

$$E_k = E_i + \frac{MV^2}{2}$$

The kinetic energy of a system thus depends on the inertial frame of reference and it is lowest with respect to the center of mass reference frame, *i.e.*, in a frame of reference in which the center of mass is stationary. In any other frame of reference there is an additional kinetic energy corresponding to the total mass moving at the speed of the center of mass.

Thus it sometimes is convenient to split the total kinetic energy of this body into the sum of the body's center-of-mass translational kinetic energy and the energy of rotation around the center of mass rotational energy:

$$E_k = E_t + E_r$$

where:

E_k is the total kinetic energy

E_t is the translational kinetic energy

E_r is the *rotational energy* or *angular kinetic energy* in the rest frame

Thus the kinetic energy of a tennis ball in flight has is the kinetic energy due to its rotation, plus the kinetic energy due to its translation.

Rotational Systems

If a rigid body is rotating, its *rotational kinetic energy* or angular kinetic energy about any line through the center of mass, and r is the distance of any mass dm from that line is simply sum of kinetic energies of its moving parts, and thus it is equal to:

$$E_r = \int \frac{v^2 dm}{2} = \int \frac{(rw)^2 dm}{2} = \frac{w^2}{2} \int r^2 dm = \frac{w^2}{2} I = \frac{1}{2} I w^2$$

where:

I is the body's moment of inertia

ω is the body's angular velocity.

The moment of inertia must be taken about an axis through the center of mass and the rotation measured by ω must be around that axis.

Relativistic Kinetic Energy of Rigid Bodies

In special relativity, we must change the expression for linear momentum. Integrating by parts, we get:

$$E_r = \int v \cdot dp =$$

$$\int v \cdot d(m\gamma v) = m\gamma v \cdot v - \int m\gamma v \cdot dv = m\gamma v^2 - \frac{m}{2} = \int \gamma d(v^2)$$

Remembering that $\gamma = (1 - v^2 / c^2)^{-1/2}$, we get:

$$E_k = m\gamma v^2 - \frac{-mc^2}{2} \int \gamma d(1 - v^2 / c^2)$$

$$= m\gamma v^2 + mc^2 (1 - v^2 / c^2)^{1/2} + C$$

And thus:

$$E_k = m\gamma\left(v^2 + c^2\left(1 - v^2 / c^2\right)\right) + C$$
$$= m\gamma\left(v^2 + c^2 - v^2\right) + C = m\gamma c^2 + C$$

The constant of integration is found by observing that $\gamma = 1$ when v = 0, so we get the usual formula:

$$E_k = m\gamma c^2 - mc^2 = \frac{mc^2}{\sqrt{1 - v^2 / c^2}} - mc^2$$

If a body's speed is a significant fraction of the speed of light, it is necessary to use relativistic mechanics (the theory of relativity as expounded by Albert Einstein) to calculate its kinetic energy.

For a relativistic object the momentum p is equal to:

$$p = \frac{mc^2}{\sqrt{1 - (v / c)^2}},$$

where *m* is the rest mass, *v* is the object's speed, and *c* is the speed of light in vacuum.

Thus the work expended accelerating an object from rest to a relativistic speed is:

$$Ek = \frac{mc^2}{\sqrt{1 - (v / c)^2}} - mc^2.$$

The equation shows that the energy of an object approaches infinity as the velocity *v* approaches the speed of light *c*, thus it is impossible to accelerate an object across this boundary. The mathematical by-product of this calculation is the mass-energy equivalence formula—the body at rest must have energy content equal to:

$$E_{rest} = mc^2$$

At a low speed (v<<c), the relativistic kinetic energy may be approximated well by the classical kinetic energy. This is done by binomial approximation. Indeed, taking Taylor expansion for square root and keeping first two terms we get:

$$E_k \approx mc^2\left(1 + \frac{1}{2} v^2 / c^2\right) - mc^2 = \frac{1}{2} mv^2.$$

So, the total energy E can be partitioned into the energy of the rest mass plus the traditional Newtonian kinetic energy at low speed.

When objects move at a speed much slower than light (*e.g.* in everyday phenomena on Earth), the first two terms of the series predominate. The next term in the approximation is small for low speeds, and can be found by extending the expansion into a Taylor series by one more term:

$$E \approx mc^2\left(1+\frac{1}{2}v^2/c^2+\frac{3}{8}v^4/c^4\right)=mc^2+\frac{1}{2}mv^2+\frac{3}{8}mv^4/c^2.$$

For example, for a speed of 10 km/s the correction to the Newtonian kinetic energy is 0.07 J/kg (on a Newtonian kinetic energy of 50 MJ/kg) and for a speed of 100 km/s it is 710 J/kg (on a Newtonian kinetic energy of 5 GJ/kg), etc.

For higher speeds, the formula for the relativistic kinetic energy is derived by simply subtracting the rest mass energy from the total energy:

$$E = m\gamma c^2 - mc^2 = mc^2\left(\frac{1}{\sqrt{1-(v/c)}}-1\right)$$

The relation between kinetic energy and momentum is more complicated in this case, and is given by the equation:

$$E_k = \sqrt{p^2c^2+m^2c^4} - mc^2$$

This can also be expanded as a Taylor series, the first term of which is the simple expression from Newtonian mechanics.

What this suggests is that the formulas for energy and momentum are not special and axiomatic, but rather concepts which emerge from the equation of mass with energy and the principles of relativity.

Quantum Mechanical Kinetic Energy of Rigid Bodies

In the realm of quantum mechanics, the expectation value of the electron kinetic energy, $\langle\hat{T}\rangle$, for a system of electrons described by the wavefunction $|\psi\rangle$ is a sum of 1-electron operator expectation values:

$$\langle\hat{T}\rangle = -\frac{\hbar^2}{2m_e}\left\langle\psi\left|\sum_{i=1}^{N}\nabla_i^2\right|\psi\right\rangle$$

where m_e is the mass of the electron and ∇_i^2 is the Laplacian operator acting upon the coordinates of the ith electron and the summation runs over all electrons. Notice that this is the quantized version of the non-relativistic expression for kinetic energy in terms of momentum:

$$E_k = \frac{p^2}{2m}$$

The density functional formalism of quantum mechanics requires knowledge of the electron density *only*, *i.e.*, it formally does not require knowledge of the wavefunction. Given an electron density $\rho(r)$, the exact N-electron kinetic energy functional is unknown; however, for the specific case of a 1-electron system, the kinetic energy can be written as

$$T(\rho)=\frac{1}{8}\int\frac{\nabla\rho(r)\cdot\nabla\rho(r)}{\rho(r)}d^3r$$

where $T[\rho]$ is known as the Von Weizsacker kinetic energy functional.

KINETIC ENERGY CORRECTION

Before claiming that this formula explains the fine structure of the hydrogen atom, however, one needs to be careful. The correction is of the order a^4, which means it of the order v^4, where v is the electron speed. The kinetic energy used in the Hamiltonian when solving the Schrφdinger equation was just $p^2/2m$, which contributed to order a^2. However, the next term in the expansion of the true relativistic kinetic energy is of order $p^4$and hence will contribute to order a^4. So if one wishes to quote the energy splittings of the hydrogen atom accurate to order a^4, one had better include the contribution from this further correction. The relativistic kinetic energy of the electron can be expanded in terms of momentum as,

$$T=\frac{p^2}{2m}-\frac{p^4}{8m^3c^2}+\ldots$$

Therefore, the correction to the Hamiltonian is,

$$\Delta H_{rel}=-\frac{1}{8m^3c^2}p^4.$$

At first sight, this looks quite complicated, since it involves the operator $p^4=\hbar^4\nabla^4$. However, one can make use of the fact that,

$$\frac{p^2}{2m}=En-V$$

to get,

$$\Delta H_{rel}=-\frac{1}{2mc^2}\left(E_n^2-2E_nV+V^2\right).$$

With V = $-ze^2/r$, applying first-order perturbation theory to this Hamiltonian reduces to the problem of finding the expectation values of r^{-1} and r^{-2}. This can be done with some effort, and the result is,

$$\Delta E_{rel}=-(Za)^4 mc^2\frac{1}{2n}\left[\frac{1}{\left(I+\frac{1}{2}\right)}-\frac{3}{4}\right].$$

Combining equations 57 and 52 and using the fact that j=l–1/2,l+1/2, the complete energy correction to order $(Za)^4$may be written

$$\Delta E_{fs} = \Delta E_{rel} + \Delta E_{so} = -(Za)^4 mc^2 \frac{1}{2n}\left[\frac{1}{\left(j+\frac{1}{2}\right)} - \frac{3}{4n}\right].$$

This energy correction depends only on j and is called the fine structure of the hydrogen atom, since it is of order $a^2 \sim 10^{-4}$times smaller than the principle energy splittings. This is why a is known as the fine-structure constant.

The fine structure of the hydrogen atom. Note that all levels are shifted down from the Bohr energies, and that for every n and l there are two states corresponding to j=l–1/2 and j=l+1/2, except for s states.

Also note that states with the same n and j but different l have the same energies, though this will be shown later not to be true due an effect know as the Lamb shift. As an aside, these fine structure splittings were derived by Sommerfeld by modifying the Bohr theory to allow elliptical orbits and then calculating the energy differences between the different states due to differences in the average velocity of the electron. By using the wrong method he got exactly the right answer, a coincidence which caused much confusion at the time. Strictly speaking, equation 58 has only been shown to be correct for $I \neq 0$ states, although it turns out to be correct for all l. To do the calculation correctly for l=0, one needs to include the effect of an additional term in the Hamiltonian known as the Darwin term, which is purely an effect of relativistic quantum mechanics and can only be understood in the context of the Dirac theory.

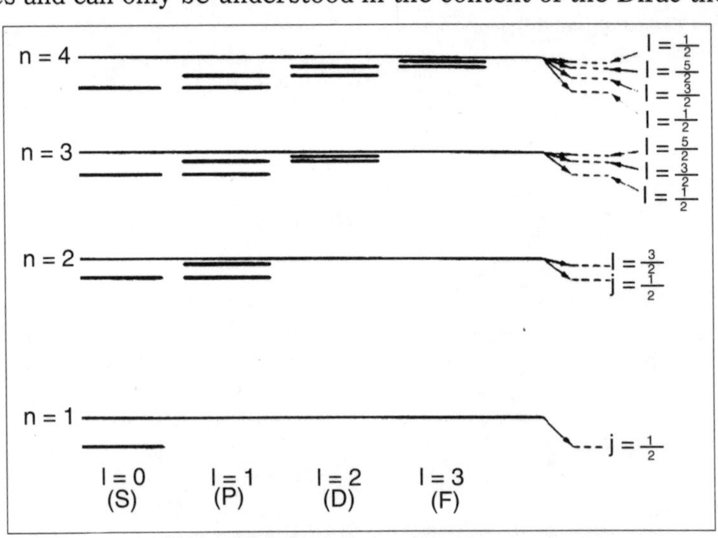

Fig. The Fine Structure of the Hydrogen Atom.
The Diagram is Not to Scale.

It is therefore appropriate to discuss the Dirac theory to achieve a more complete understanding of the fine structure of the hydrogen atom.

LAMINAR AND TURBULENT FLOW IN FLUID DYNAMICS

The nature of the flow of a fluid by injecting a fine filament of dye into the stream of flow and taking note of what happens to this filament. It was found in experiments that at low velocities the dye filament remained intact and that the filaments made parallel lines in the stream of flow. This is known as Laminar flow (or viscous or streamline). If the velocity of flow is gradually increased, the dye filament is eventually broken up and spread over the cross section of the pipe. This is turbulent flow, in which the particles of fluid are not moving in parallel lines but are moving across the general direction of flow.

If a fluid particle in a stream is disturbed, its inertia will tend to move it in a new direction, however the viscous forces from the surrounding fluid will tend to move it in the general direction of flow. If the shear forces are large enough to overcome any deviation, then we have viscous or laminar flow. However, if the shear forces are relatively weaker, and not sufficient to overcome the inertia of the particles, then we have turbulent flow.

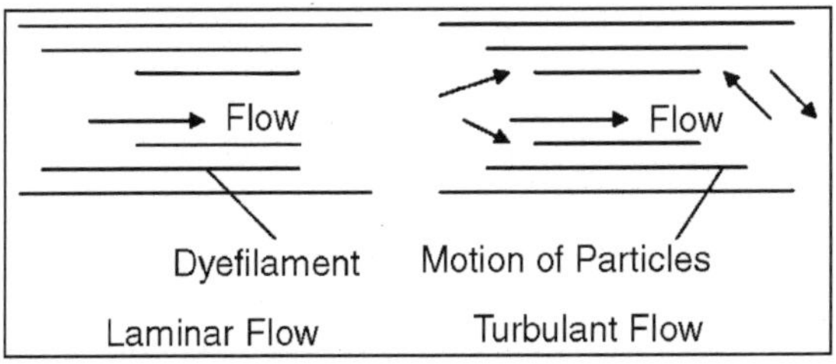

Therefore it is the ratio of the inertia to the viscous forces which determines whether flow will be laminar or turbulent.

The ratio of the inertia forces to the viscour forces is given by:

$$\rho\frac{ul}{\mu} \text{ (Reynolds Number)}$$

Therefore, it is the Reynolds number which determines whether a flow will be laminar or turbulent.

As Kinematic Viscosity $(\nu) = \frac{\mu}{\rho}$,

Reynolds number can also be expressed as cl νl is the characteristic dimension in the system *e.g.* the diameter of the pipe in which the fluid is flowing.

Experiments have shown that if the Reynolds number is less than 2000, then the flow will be laminar. If it is above 2000 it is likely to be turbulent, however turbulent flow can exist at much higher Reynolds numbers.

Example: Water flows through a pipe 30mm diameter at a velocity of 10 m/s. If the dynamic viscosity of water is taken as 1.30×10^{-3} kg/ms and its density as 1000 kg/m^3, determine whether the flow will be laminar or turbulent.

Reynolds number,

$$= \rho \frac{c1}{\mu}$$

$$= \frac{1000 \times 10 \times 0.03}{1.3 \times 10^{-3}}$$

$$= 230769$$

This is significantly larger than 2000 and therefore the flow will be turbulent.

Problem If oil of specific gravity 0.85 and dynamic viscosity 14×10^{-2} kg/ms is pumped through the pipe at the same viscocity, what type of flow will occur. If we were to take a pipe of free flowing water and inject a dye into the middle of the stream, what would we expect to happen?

This,

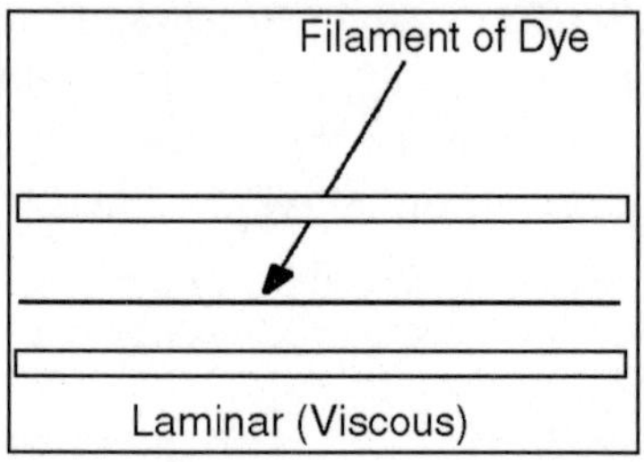

this,

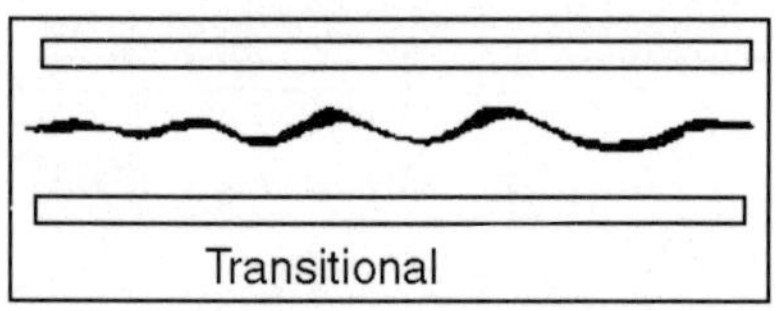

or this

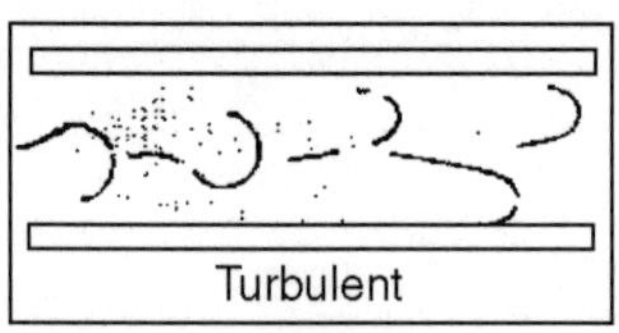

Actually both would happen - but for different flow rates. The top occurs when the fluid is flowing fast and the lower when it is flowing slowly.

The top situation is known as turbulent flow and the lower as laminar flow. In laminar flow the motion of the particles of fluid is very orderly with all particles moving in straight lines parallel to the pipe walls.But what is fast or slow? And at what speed does the flow pattern change? And why might we want to know this? The phenomenon was first investigated in the 1880s by Osbourne Reynolds in an experiment which has become a classic in fluid mechanics.

$$\frac{pud}{\mu}$$

He used a tank arranged as above with a pipe taking water from the centre into which he injected a dye through a needle. After many experiments he saw that this expression

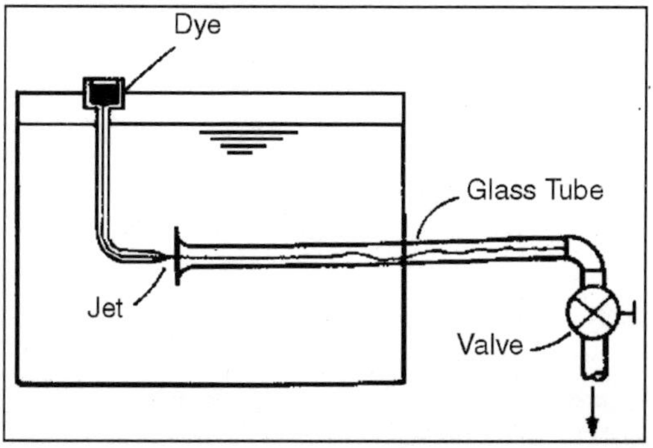

where ρ = density, u = mean velocity, d = diameter and μ = viscosity would help predict the change in flow type. If the value is less than about 2000 then flow is laminar, if greater than 4000 then turbulent and in between these then in the transition zone.

This value is known as the Reynolds number, Re:

$$\text{Re} = \frac{pud}{\mu}$$

Laminar flow: Re < 2000
Transitional flow: 2000 < Re < 4000
Turbulent flow: Re > 4000

Units of Reynolds number

We can fill in the equation with SI units:

$\rho = \text{kg/m}^3$, $u = \text{m/s}$, $d = \text{m}$

$\mu = \text{Ns/M}^2 = \text{kg/ms}$

$$\text{Re} = \frac{pud}{\mu} = \frac{\text{kg}}{\text{m}^3}\frac{\text{m}}{\text{s}}\frac{\text{m}}{1}\frac{\text{m}}{\text{kg}} = 1$$

i.e. it has no units. A quantity that has no units is known as a non-dimensional (or dimensionless) quantity. Thus the Reynolds number, Re, is a non-dimensional number.

We can go through an example to discover at what velocity the flow in a pipe stops being laminar.

If the pipe and the fluid have the following properties:
water density ρ = 1000 kg/m^3
pipe diameter d = 0.5m
(dynamic) viscosity, μ = 0.55x10^3 Ns/m^2
We want to know the maximum velocity when the Re is 2000.

$$\text{Re} = \frac{\text{pud}}{\mu} = 2000$$

$$\mu = \frac{2000\mu}{\text{pd}} = \frac{2000 \times 0.55 \times 10^{-3}}{1000 \times 0.5}$$

$$\text{m} = 0.0022 \text{ m/s}$$

If this were a pipe in a house central heating system, where the pipe diameter is typically 0.015m, the limiting velocity for laminar flow would be, 0.0733 m/s. Both of these are very slow. In practice it very rarely occurs in a piped water system - the velocities of flow are much greater. Laminar flow does occur in situations with fluids of greater viscosity - *e.g.* in bearing with oil as the lubricant. At small values of Re above 2000 the flow exhibits small instabilities. At values of about 4000 we can say that the flow is truly turbulent. Over the past 100 years since this experiment, numerous more experiments have shown this phenomenon of limits of Re for many different Newtonian fluids - including gasses.

Abstract Number Mean

We can say that the number has a physical meaning, by doing so it helps to understand some of the reasons for the changes from laminar to turbulent flow.

$$\text{Re} = \frac{\text{pud}}{\mu}$$

$$= \frac{\text{inertial forces}}{\text{viscous forces}}$$

It can be interpreted that when the inertial forces dominate over the viscous forces (when the fluid is flowing faster and Re is larger) then the flow is turbulent. When the viscous forces are dominant (slow flow, low Re) they are sufficient enough to keep all the fluid particles in line, then the flow is laminar.

LAMINAR FLOW

Laminar flow is also referred to as streamline or viscous flow. These terms are descriptive ofthe flow because, in laminar flow:

- Layers of water flowing over one another at differentspeeds with virtually no mixing between layers,
- Fluid particles move in definite andobservable paths or streamlines.
- The flow is characteristic of viscous (thick) fluid or isone in which viscosity of the fluid plays a significant part.
  - Re < 2000
  - 'low' velocity
  - Dye does not mix with water
  - Fluid particles move in straight lines
  - Simple mathematical analysis possible
  - Rare in practice in water systems.

TURBULENT FLOW

Turbulent flow is characterized by the irregular movement of particles of the fluid. There is nodefinite frequency as there is in wave motion. The particles travel in irregular paths with noobservable pattern and no definite layers.

- Re > 4000
- 'high' velocity
- Dye mixes rapidly and completely
- Particle paths completely irregular
- Average motion is in the direction of the flow
- Cannot be seen by the naked eye
- Changes/fluctuations are very difficult to detect. Must use laser.
- Mathematical analysis very difficult - so experimental measures are used
- Most common type of flow.

PRESSURE LOSS

Up to this point on the course we have considered ideal fluids where there have been no losses due to friction or any other factors. In reality, because fluids are viscous, energy is lost by flowing fluids due to friction which must be taken into account. The effect of the friction shows itself as a pressure (or head) loss.

In a pipe with a real fluid flowing, at the wall there is a shearing stress retarding the flow, as shown below.

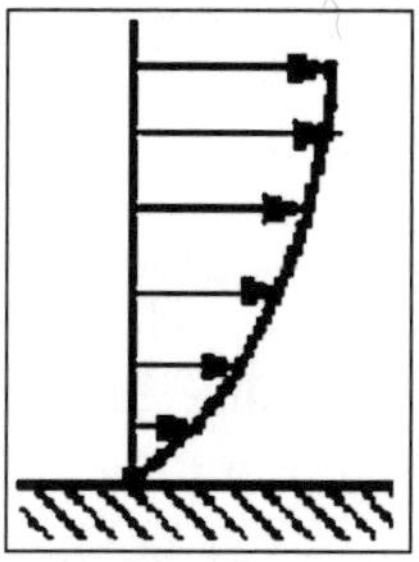

If a manometer is attached as the pressure (head) difference due to the energy lost by the fluid overcoming the shear stress can be easily seen. The pressure at 1 (upstream) is higher than the pressure at 2.

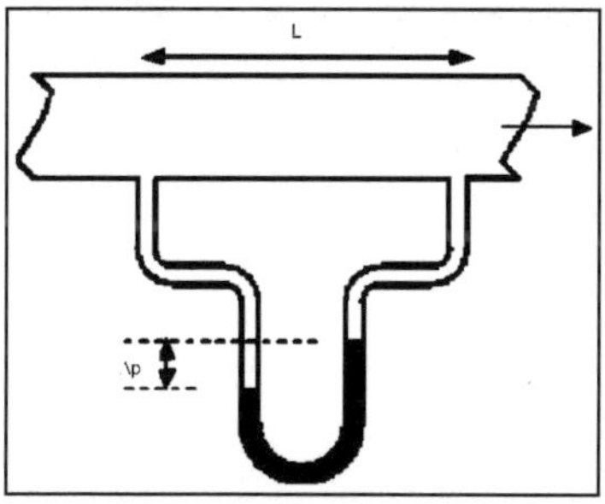

We can do some analysis to express this loss in pressure in terms of the forces acting on the fluid. Consider a cylindrical element of incompressible fluid flowing in the pipe, as shown,

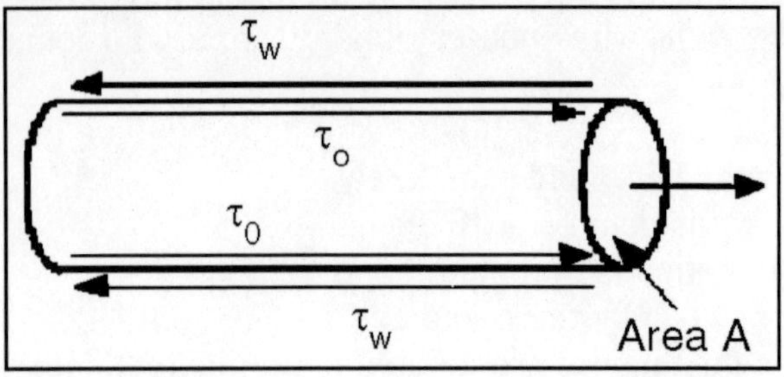

The pressure at the upstream end is *p*, and at the downstream end the pressure has fallen by Δp to $(p-\Delta p)$. The driving force due to pressure (F = Pressure x Area) can then be written driving force = Pressure force at 1 - pressure force at 2,

$$pA-(p-\Delta p)A=\Delta p A=\Delta p\frac{\pi d^2}{4}$$

The retarding force is that due to the shear stress by the walls,

= shear stress × area over which it acts

= τ_w × area of pipe wall

= $\tau_w \pi dL$

As the flow is in equilibrium, driving force = retarding force,

$$\Delta p\frac{\pi d^2}{4}=\tau_w \pi dL$$

$$\Delta p=\frac{\tau_{w4L}}{d}$$

Giving an expression for pressure loss in a pipe in terms of the pipe diameter and the shear stress at the wall on the pipe.

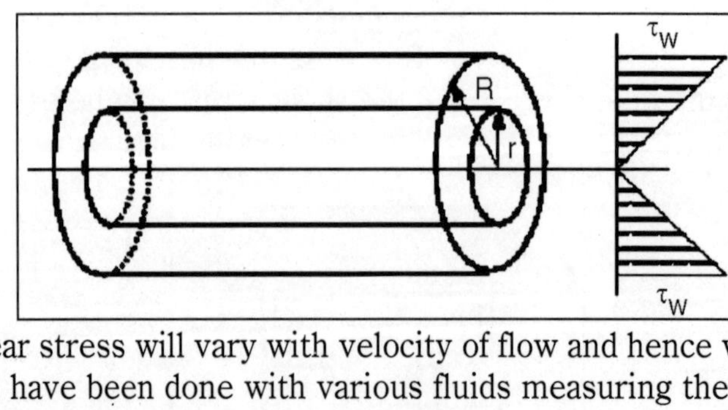

The shear stress will vary with velocity of flow and hence with Re. Many experiments have been done with various fluids measuring the pressure loss at various Reynolds numbers. These results plotted to show a graph of the relationship between pressure loss and Re look similar to the figure below:

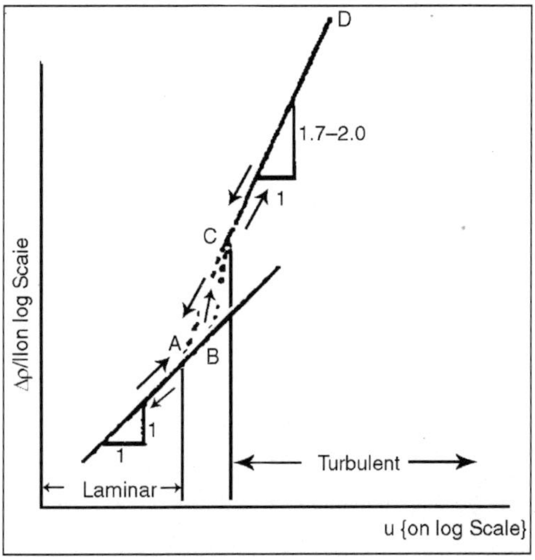

This graph shows that the relationship between pressure loss and Re can be expressed as

laminar $\Delta p \propto \mu$

turbulent $\Delta p \propto \mu^{17 (\text{or } 2.0)}$

As these are empirical relationships, they help in determining the pressure loss but not in finding the magnitude of the shear stress at the wall τ_w on a particular fluid. If we knew τ_w we could then use it to give a general equation to predict the pressure loss.

PRESSURE LOSS DURING LAMINAR FLOW

In general the shear stress τ_w. is almost impossible to measure. But for laminar flow it is possible to calculate a theoretical value for a given velocity, fluid and pipe dimension. In laminar flow the paths of individual particles of fluid do not cross, so the flow may be considered as a series of concentric cylinders sliding over each other - rather like the cylinders of a collapsible pocket telescope.

As before, consider a cylinder of fluid, length L, radius r, flowing steadily in the centre of a pipe.

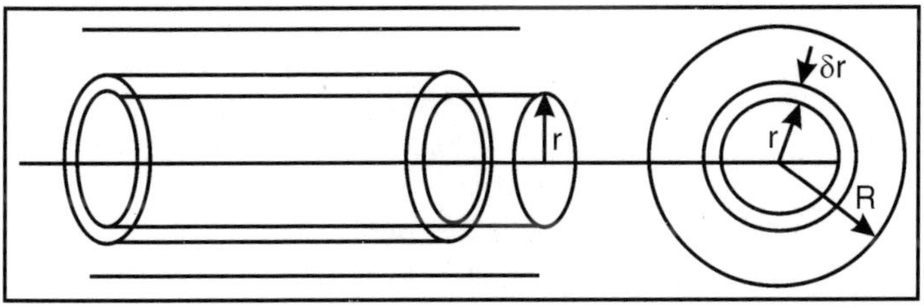

We are in equilibrium, so the shearing forces on the cylinder equal the pressure forces.

$$\tau\, 2\pi r L = \Delta p A = \pi r^2$$

$$\tau = \frac{\Delta p}{L}\frac{r}{2}$$

By Newtons law of viscosity we have,

$$\tau = \mu\frac{du}{dy},$$

where y is the distance from the wall. As we are measuring from the pipe centre then we change the sign and replace y with r distance from the centre, giving,

$$\tau = \mu\frac{du}{dr}$$

Which can be combined with the equation above to give,

$$\frac{\Delta p}{L}\frac{r}{2} = \mu\frac{du}{dr}$$

$$\frac{du}{dr} = -\frac{\Delta p}{L}\frac{r}{2\mu}$$

In an integral form this gives an expression for velocity,

$$u = -\frac{\Delta p}{L}\frac{1}{2\mu}\int r\,dr$$

Integrating gives the value of velocity at a point distance r from the centre,

$$u_r = -\frac{\Delta p}{L}\frac{r^2}{4\mu} + C$$

At $r = 0$, (the centre of the pipe), $u = u_{max}$, at $r = R$ (the pipe wall) $u = 0$, giving,

$$C = \frac{\Delta P}{L}\frac{R^2}{4\mu}$$

so, an expression for velocity at a point r from the pipe centre when the flow is laminar is,

$$u_r = \frac{\Delta p}{L}\frac{1}{4\mu}(R^2 - r^2)$$

Note how this is a parabolic profile (of the form $y = ax^2 + b$) so the velocity profile in the pipe looks similar to the figure below,

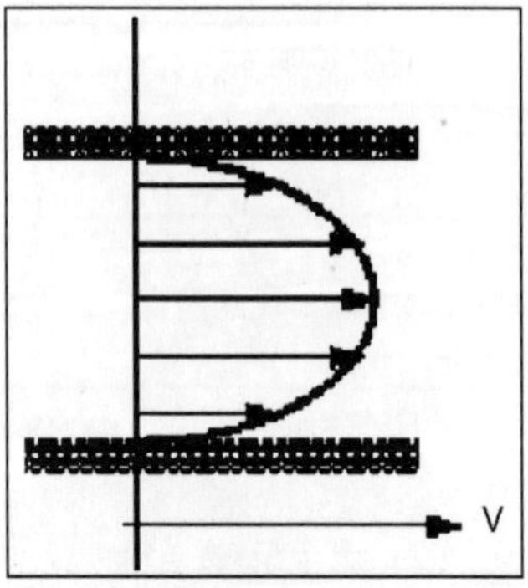

What is the discharge in the pipe?

$$Q = u_m A$$

$$u_m = \int_0^R u_r, dr$$

$$= \frac{\Delta p}{L}\frac{1}{4\mu}\int_0^R u_r\, dr$$

$$= \frac{\Delta p}{L}\frac{R^2}{8\mu} = \frac{\Delta p d^2}{32\mu L}$$

So the discharge can be written,

$$Q = \frac{\Delta p d^2}{32\mu L}\frac{\pi d^2}{4}$$

$$= \frac{\Delta p}{L}\frac{\pi d^2}{128\mu}$$

This is the Hagen-Poiseuille equation for laminar flow in a pipe. It expresses the discharge Q in terms of the pressure gradient,

$$\left(\frac{\partial p}{dx} = \frac{\Delta p}{L}\right),$$

diameter of the pipe and the viscosity of the fluid.

We are interested in the pressure loss (head loss) and want to relate this to the velocity of the flow. Writing pressure loss in terms of head loss h_f, *i.e.* $p = rgh_f$

$$u = \frac{pgh_f d^2}{32\mu L}$$

$$h_f = \frac{32\mu L u}{pgd^2}$$

This shows that pressure loss is directly proportional to the velocity when flow is laminar.

It has been validated many time by experiment.

It justifies two assumptions:

- Fluid does not slip past a solid boundary
- Newtons hypothesis.

7

Hydraulic Cylinder

Available in a gamut of sizes, types, and design configurations, cylinder actuators form the main output device of a hydraulic motion system, and often the most visible part. These actuators convert hydraulic fluid pressure into rapid, controllable linear motion and force to move loads.

A typical actuator consists of the cylinder body, end caps, piston, piston rod, seals, and bearing surfaces for piston and rod.

Designed for various pressures, ratings up to 3,000 psi (continuous) apply to industrial units; up to 5,000 psi for mill and press type units. Stroke lengths to 10 ft and cylinder diameters to 8 in. are available, with much larger units made for special applications.

From a basic relationship of hydraulics (Pascal's Law), linear force magnitude developed by a cylinder is the system fluid pressure times the piston's *effective area* over which pressure acts, or $F = PA$. Friction and other real-world losses will reduce results from this simple relation.

MANY CONFIGURATIONS

The simplest actuator configuration is a *single-acting cylinder*, where fluid is ported to one side of the piston, producing output force and motion only in one direction. Gravity or external springs return the piston to its starting position as the fluid is returned to the reservoir. A *double-acting cylinder* ports fluid to either side of the piston to produce force/motion when extending and retracting the piston rod. Sealing between piston OD and cylinder ID must handle both directions of motion. A variant here is a *double-end rod cylinder* that adds a piston rod extending through the cylinder's rear end cap.

Fig. Typical industrial hydraulic cylinder

The double-acting cylinder is the most common hydraulic cylinder type, providing output stroke in two directions. Industrial cylinders favour tie-rod construction to lock cylinder body, head, and end cap in place.

A typical double-acting cylinder produces somewhat greater force when extending than when retracting—with equal pressure applied on each side of the piston—due to different effective areas exposed to the hydraulic fluid. The rod-side piston area is smaller than the head side, resulting in less force on the retract stroke, says Pascal's Law. A double-end rod cylinder has essentially the same area on each side of the piston, and doesn't exhibit this *differential cylinder* effect.

A variant of the above actuator types is a *ram cylinder*, designed with a robust diameter piston rod approaching full-piston diameter that prevents rod bending in a long-stroke horizontal unit or buckling under high vertical loads as in a press or stamping application. The piston connects to a high-strength steel piston rod whose opposite end comes with various rod-end connections. Surface hardened or chrome-plated rods are often used with ultra-fine surface finish that's key to long seal life.

Piston OD must fit precisely to the cylinder ID. Each must be precisely cylindrical and extremely finely finished for smooth output motion.

The latest industrial cylinders can incorporate sensor feedback and an electrohydraulic servo valve for sophisticated speed and position control of loads—not only full stroke (in/out) or rough positioning. Applications range from production machine tools and material handling to steel mill actuators, nuclear power plant controls, and passenger and freight elevators.

Hydraulic Cylinder Controls, Applications

As mentioned in the main article, control electronics associated with hydraulic cylinders help increase the performance of these devices. At the high end are electrohydraulic proportional (servo) valves and sophisticated feedback sensors integrated into the cylinder that challenge electric drives for certain applications. More common are directional control valves that actuate cylinders automatically or manually.

Hydraulic cylinders work in numerous applications. A large sector is "mobile hydraulics," where hydraulic systems power various motions and tools for off-road vehicles and equipment.

Here, a reversing directional valve (single- or double-acting, depending on cylinder type) typically controls hydraulic fluid flow control to and from the cylinder.

Differential cylinders produce greater force in extension than in retraction.

However, if it's important to have equal force output in each direction, the rod-side supply pressure could be increased to make that adjustment. Another effect of the differential cylinder is faster speed in retraction than extension, which seems counterintuitive at first. It's explained by less volume capacity on the cylinder's rod-side allowing for faster retraction—with equal fluid flow going to both ends of the cylinder. In the U.S., numerous standards of the National Fluid Power Association (NFPA) are widely applied in the design and manufacture of hydraulic cylinders.

WATER HYDRAULICS

While hydraulic cylinders operate predominantly on pressurized hydraulic oil, a niche technology exists—called "water hydraulics"—that applies pure tap water at high pressures as the "hydraulic fluid" to power cylinders, actuators, and other devices.

Water hydraulics has a substantial following in Europe, but various products and systems are available elsewhere. One notable supplier of the technology is the Danish company Danfoss A/S, with its high-pressure water solution trademarked "Nessie."

More Cylinder Types

A multi-stage telescoping cylinder delivers increased stroke length from a cylinder body of given length. Total stroke can be about four times the retracted length of the unit, with either single- or double-acting cylinder styles available. Output force is highest with the first stage and decreases as each additional stage extends. These cylinders are often of the "ram" type. Telescoping cylinders are commonly used in mobile hydraulic systems.

A *cushioned cylinder* refers to the incorporation of a device in the actuator to minimize shock loads as the piston comes to the end(s) of its stroke.

Cushioning can be accommodated at one or both ends of the cylinder typically by means of a metering device that restricts flow at the applicable cylinder outlet port(s) to slow the piston's motion at end of stroke.

Certain manufacturing applications require a short-stroke cylinder, offered by various suppliers, that provide shorter piston strokes and smaller body sizes than standard cylinders. One manufacturer of these actuators, Mack Corp., offers an extensive number of basic cylinder configurations plus additional custom designs. The company offers a free wall chart describing 88 basic cylinder configurations it offers. Support for dynamic stage and cinema productions is among the wide applications of hydraulic technology.

HYDROSTATIC PARADOX

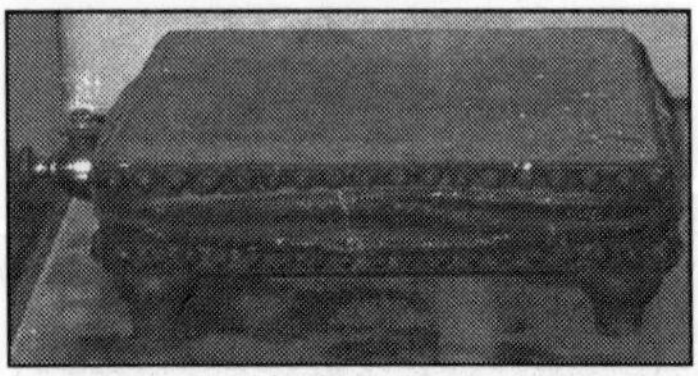

The hydrostatic paradox arises from our failure to accept, at first sight, the conclusion published by Blaise Pascal in 1663: the pressure at a certain level in a fluid is proportional to the *vertical* distance to the surface of the liquid. Pascal's Vases are well known illustrations of the paradox, but the two demonstrations on this page are little known. The hydrostatic bellows apparatus above (at Transylvania University) and to the right (from the 1871 edition of Denison Olmsted's Natural Philosophy text) were sold by many 19th century equipment manufacturers. The 1860 Ritchie catalogue offered this model "of improved form and construction; mahogany, twelve inches square, with patent leather sides lined with vulcanized rubber; brass socket and three-way water cock; brass and glass tubes with brass screw connections; funnel...$10.00"

It must have been a great surprise to see a 150 pound student supported by a column of water reaching only 2.4 ft above the level of the platform.

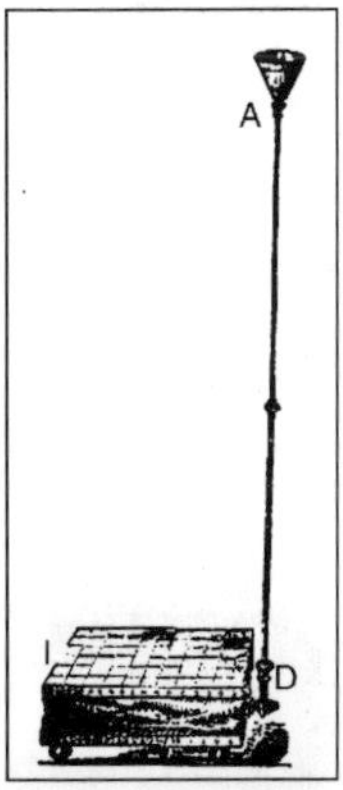

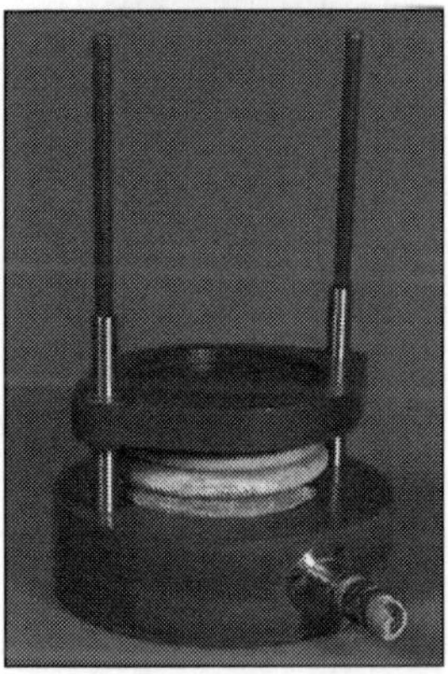

The apparatus at the left is is "Ritchie's Illustration of the Hydraulic Press; mahogany base with a sliding platform supported by brass pillars, with a weight; upon the base is placed a bellows-shaped rubber bag, connected by a tube to a globular bag fitted with a cap and cork.

Fill the globe with water, and elevate it; the pressure of the column will force the water into the bellows, raising the weight; lower the globe, and the weight will force the water back into it... $12.00".

Middlebury College's surviving example of this apparatus is in remarkably good condition after one hundred and more years. It needs only the replacement of rubber parts to make it work once more.

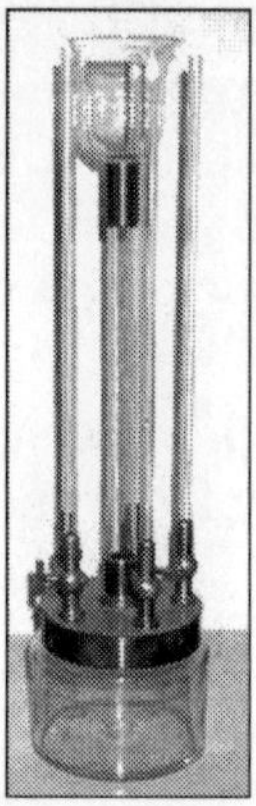

This version of the hydrostatic paradox was in storage at the museum at the United States Military Academy at West Point in February 2000.

The same apparatus, shown in below was on display in the Garland Collection of Classical Physics Apparatus at Vanderbilt University when I visited it in March 2001.

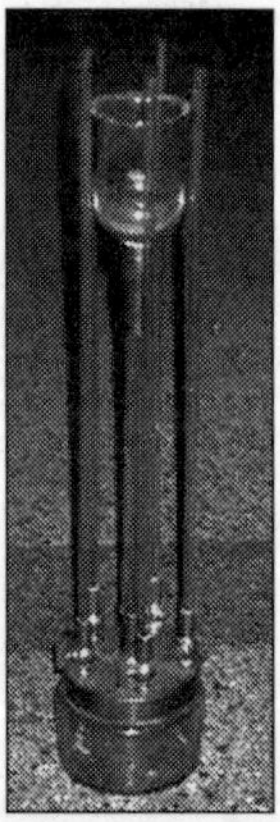

The apparatus is listed in the 1865 Deleuil catalogue as a device to show that all liquid-filled columns exert pressure independently of each other. The apparatus cost 40 francs, or $10.00. We would fill each column with a liquid of a different density, with water filling the central column and reservoir. The taps would then be slowly opened, and liquids added to keep the interfaces at the same level. The Millington./Barnard Collection at the University of Mississippi Museum has an example of the Hydrostatic Paradonx identical to the ones at West Point and Vandcerbilt.

BUOYANCY

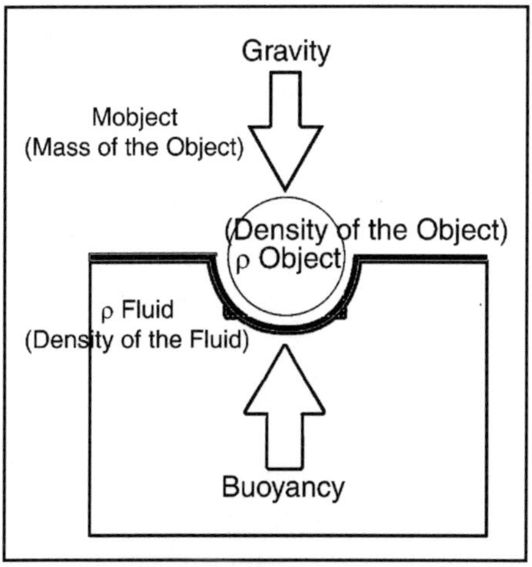

In physics, buoyancy is the upward force on an object produced by the surrounding fluid (*i.e.*, a liquid or a gas) in which it is fully or partially immersed, due to the pressure difference of the fluid between the top and bottom of the object. The net upward buoyancy force is equal to the magnitude of the weight of fluid displaced by the body. This force enables the object to float or at least to seem lighter. Buoyancy is important for many vehicles such as boats, ships, balloons, and airships.

Buoyancy acts against the force of gravity and so makes objects seem lighter with respect to gravity. To represent this effect, which is important for sedimentation, it is common to define a *buoyant mass* m_b that represents the effective mass of the object with respect to gravity,

$$m_b = m_{object} \cdot \left(1 - \frac{\rho fluid}{\rho object}\right)$$

where m_{object} is the true (vacuum) mass of the object, whereas ρ_{object} and ρ_{fluid} are the average densities of the object and the surrounding fluid, respectively. Thus, if the two densities are equal, $\rho_{object} = \rho_{fluid}$, the object appears to be weightless. If the fluid density is greater than the average density of the object, the object floats; if less, the object sinks.

Forces and Equilibrium

Buoyancy provides an upward force on the object. The magnitude of this force is equal to the weight of the displaced fluid. (Displacement is the term used for the weight of the displaced fluid and, thus, is an equivalent term to buoyancy.) The buoyancy of an object depends, therefore, only upon two factors: the object's submerged volume, and the density of the surrounding fluid. The greater the object's volume and surrounding density of the fluid, the more buoyant force it experiences. If the buoyancy of an (unrestrained and unpowered) object exceeds its weight, it tends to rise. An object whose weight exceeds its buoyancy tends to sink.

The atmosphere's density depends upon altitude. As an airship rises in the atmosphere, therefore, its buoyancy reduces as the density of the surrounding air reduces. The density of water is essentially constant: as a submarine expels water from its buoyancy tanks (by pumping them full of air) it rises because its volume stays the same (the volume of water it displaces if it is fully submerged) while its weight is decreased.

The buoyant force can be expressed using the following equation:

$$F_{buoyant} = -\rho Vg$$

where,

- ρ is the density of the fluid;
- V is the volume of the object submerged;
- g is the standard gravity ($\approx$ 9.81 N/kg on Earth). A negative sign must be used because the buoyancy acts in the opposite direction to the acceleration due to gravity.

COMPRESSIBLE OBJECTS

As a floating object rises or falls the forces external to it change and, as all objects are compressible to some extent or another, so does the object's volume. Buoyancy depends on volume and so an object's buoyancy reduces if it is compressed and increases if it expands. If an object at equilibrium has a compressibility less than that of the surrounding fluid, the object's equilibrium is stable and it remains at rest.

If, however, its compressibility is greater, its equilibrium is then unstable, and it rises and expands on the slightest upward perturbation, or falls and compresses on the slightest downward perturbation. A submarine is more compressible than the surrounding water. As depth increases, the resulting pressure causes the submarine's volume to decrease more than the volume of the surrounding water decreases. Buoyancy depends upon the object's volume and the weight of the displaced fluid. Volume has decreased so the weight displaced has decreased which means a decrease in buoyancy and the submarine tends to sink further. A rising submarine expands more than the surrounding water, so tends to rise further.

The height of a balloon tends to be stable. As a balloon rises it tends to increase in volume with reducing atmospheric pressure, but the balloon's cargo does not expand. The average density of the balloon decreases less, therefore, than that of the surrounding air. The balloon's buoyancy reduces because the weight of the displaced air is reduced. A rising balloon tends to stop rising. Similarly a sinking balloon tends to stop sinking.

SHOCK WAVES AND CONTINUUM FLOW MODELS

Since the conservation equation describes flow and density as a function of distance and time, one can immediately see that continuum modeling is superior to input-output models used in practice (which are only one dimensional, because they essentially ignore space). In addition, because flow is assumed to be a function of density, continuum models have a second major advantage, (*e.g.* compressibility).

The simple continuum model referred to in this text consists of the conservation equation and the equation of state (speed-density or flow density relationship). If these equations are solved together with the basic traffic flow equation (flow equals density times speed), then we can obtain speed, flow, density at any time and point of the roadway.

Knowing these basic traffic flow variables we know the state of the traffic system and can derive measures of effectiveness, such as delays stops, total travel, travel time, and others that allow engineers to evaluate how well the system is performing....

A shock wave is a discontinuity of flow or density, and has the physical implication that cars change speeds abruptly without time to accelerate or decelerate. This is an unnatural behaviour that could be eliminated by considering high order continuum models.

These models add a momentum equation that accounts for the acceleration and inertia characteristics of traffic mass. In this manner, shock waves are smoothed out and the equilibrium assumption is removed.... In spite of this improvement, the most widely known high order models still require an equilibrium speed-density relationship....

$$\left[\mathrm{uf} - 2\mathrm{uf}\frac{\mathrm{k}}{\mathrm{kj}}\right]\frac{\partial \mathrm{k}}{\partial \mathrm{x}} + \frac{\partial \mathrm{k}}{\partial \mathrm{t}} = 0$$

where,

u_f represents the free flow speed

k_j the jam density is a first order quasi-linear, partial differential equation which can be solved by the method of characteristics.

In practical terms, the solution suggests that:

- The density k is a constant along a family of curves called characteristics or waves; a wave represents the motion (propagation) of a change in flow and density along the roadway.
- The characteristics are straight lines emanating from the boundaries of the time-space domain.
- *The slope of the characteristics is*:

$$\frac{\mathrm{dx}}{\mathrm{dt}} = \mathrm{f(k)} + \mathrm{k}\left[\mathrm{f(k)}\right] = \frac{\mathrm{dq}}{\mathrm{dk}}$$

- This implies that the characteristics have slope equal to the tangent of the flow-density curve at the point representing the flow conditions at the boundary from which the characteristic emanates.

- The density at any point x,t of the time space domain is found by drawing the proper characteristic passing through that point.
- The characteristics carry the value of density (and flow) at the boundary from which they emanate.
- When two characteristic lines intersect, then density at this point should have two values which is physically unrealizable; this discrepancy is explained by the generation of shock waves. In short, when two characteristics intersect, a shock wave is generated and the characteristics terminate. A shock then represents a mathematical discontinuity (abrupt change) in k, q, or$u$.

The speed of the shock wave is:

$$u_w = \frac{q_d - q_u}{k_d - k_u}$$

where k_d, q_d represent downstream and k_u, q_u upstream flow conditions. In the flow concentration curve, the shock wave speed is represented by the slope of the line connecting the two flow conditions (*i.e.*, upstream and downstream).

It should be noted that when u_w is positive, the shock wave moves downstream with respect to the roadway; conversely, when u_w is negative, the shock is moving upstream.

Furthermore, the mere fact that a difference exists in flow conditions upstream and downstream of a point does not imply that a shock wave is present unless the characteristics intersect.

Generally this occurs only when the downstream density is higher than upstream. When density downstream is lower than upstream, we have diffusion of flow similar to that observed when a queue is discharging. When downstream density is higher than upstream, then shock waves are generated and queues are generally being built even though they might be moving downstream.

Figure demonstrates the use of traffic waves in identifying the occurrence of a shock wave and following its trajectory. The process follows the steps of the solution of the conservation equation as outlined above. The top of the figure represents a glow-concentration curve; the bottom figure represents trajectories of the traffic waves. On the *q-k* curve, point A represents a situation where traffic flows at near capacity implying that speed is well below the free-flow speed.

Point B represents an uncongested condition where traffic flows at a higher speed because of the lower density. Tangents at points A and B represent the wave velocities of these two situations. The areas where conditions A and B prevail are shown by the characteristics drawn in the bottom. This figure assumes that the faster flow of point B occurs later in time than that of point A; therefore, the characteristics (waves) of point B will eventually intersect with those of point A. The intersection of these two sets of waves has a slope equal to the chord connecting the two points on the *q-k* curve, and this intersection represents the path of the shock wave shown at the bottom.

It is necessary to clarify that the waves of the time-space are not the trajectories of vehicles but lines of constant flow and speed showing the propagation of conditions A and B. The velocities of individual vehicles within A and B are higher because the speed of the traffic stream is represented by the line connecting the origin with A and B in the *q-k* curve.

FREE MOLECULAR FLOW

Definition

A flow regime in aerodynamics in which molecules emitted from an object, as it passes through a resistive medium, do not affect the flow of oncoming molecules by scattering interactions, *i.e.*, the mean free path of the emitted molecules in much longer than a characteristic linear dimension of an object.2. Flow about a body in which the number of collisions between the molecules of the fluid is negligible compared with the collisions between these molecules and the body. Also called free molecular flow. See rarefied gas dynamics

The transmission probabilities and the number of molecules reflecting from the wall were computed using a Monte Carlo method for free molecule flow through a conduit containing an orifice of zero thickness. The computed transmission probabilities were compared with a composite transmission rule and a series-resistance equation.

The distribution of the molecular number density on the conduit wall along the conduit axis was determined by assuming that the ratio of the molecular number flux reflecting from the conduit wall to that entering into the conduit inlet corresponds to the ratio of the molecular number densities at each position.

When the orifice was placed halfway down the length of the conduit, the composite rule and the series-resistance equation, respectively, agreed well with the computed transmission probabilities for the conduits with the aspect ratio $L/R \geq 10$, and the cross-sectional area ratio of tube and orifice opening $\beta \geq 10$, and for the conduit with $\beta \geq 40$.

The molecular number density distributions were symmetric with respect to the midpoint, and had a minimum in the upstream side and a maximum in the downstream side. The molecular number density distributions exhibited jumps across the orifice, as well as at the inlet and exit of the conduit.

Gas-flow phenomenon at low pressures or in small channels when the mean free path is of the same order of magnitude as the channel diameter; a gas molecule thus migrates along the channel independent of other gas molecules present.

8

Stability of Floating Bodies

STABILITY

Stability is a measure of the tendency of an ocean vehicle to return to upright if inclined by an external force. For different condition of operations, stability can be classified into the following categories: Intact stability (static stability and dynamic stability) and damage stability. It is important to know that the inclination force acting upon it shall not affect its safety.

Purpose

The purposes of this experiment are:

- To determine experimentally, for a body floating in water, the metacentric height and its centre gravity.
- To investigate the effect of adding a weight vertically above the C.G. on the stability of a floating body.
- To investigate the free surface effect on the stability of a floating body.

Scope

In this experiment, only intact static stability of a vessel at small angle of heel (f < 10°) is considered. The theories of operation and the procedure adopted to evaluate the stability qualities are given in this manual.

DESCRIPTION OF EQUIPMENT

The equipment consists of a rectangular barge with water-tight sub-division compartments. The inclining moment is provided by weights which can be moved transversely on either side of the amidship section of the barge, of which the distance from the centre-line being measureable. The angle of heel is taken by means of a precision clinometer.

Theory of Operation

- Consider a floating body being inclined at a small angle f by an external force.

$$\text{Righting moment} = \Delta\overline{CZ} = \Delta\overline{GM_T\phi}$$

- If G is below M_T, the vessel is in stable equilibrium
- If G is above M_T, the vessel is in unstable equilibrium
- If G concedes with M_T, the vessel is in neutral equilibrium

It may be shown that, period of oscillation for small angle of displacement of the above mentioned floating body is,

$$T = 2\pi\left(\frac{Kxx}{\sqrt{g\,GM_T}}\right)$$

where Kxx is the radius of gyration about the longitudinal axis.

Thus the greater is GM_T, the more rapid the oscillation. The above are two conflicting requirements. A good design should provide adequate but not excessive GM_T

- *Inclined experiment for finding GM_T and KG*: In this experiment, mass m is moved transversely across the deck through a known distance d. For small angle of heel.

$$\therefore GM_T = \frac{mdg}{\Delta \text{Tan}\phi}$$

$$\Delta = \text{weight of water displaced} = \Delta\rho s g$$

$$\overline{KG} = \overline{KB} + \overline{BM_T} - \overline{GM_T}$$

where,

$$\text{BMT} = \frac{1}{\nabla},$$

$$\nabla = \text{volume of water}, \nabla = \rho s \nabla \text{ displaced}$$

and $\overline{KB}$ is obtained from hydrostatic calculation.

$$\overline{KB} = \frac{1}{2}(\text{Draught})$$

- *Effects of Free surfaces*: When a tank on board an ocean vehicle is notcompletely full'Of liquid (Fi-ur,- 4), the liquid will move across the tank in the same direction as the vessel durino, rolling.

The centre of gravity of the vessel will move away from the centre-line plane. Then the righting arm will be reduced.

Virtual reduction in Metacentiic height due to effect of free surface is,

$$GG_2 = \frac{\rho_1}{\rho_s}\frac{\Sigma_i}{\nabla}$$

where i is the second moment of area of the free surface of the tank about its centre line.

Hence, the effective Metacentric Height is,

$$\overline{G_2M_T} = \overline{GM_T} - \overline{GG_2}$$

FLOATING BODY

A floating body is STABLE if, when it is displaced, it returns to equilibrium.

A floating body is UNSTABLE if, when it is displaced, it moves to a new equilibrium.

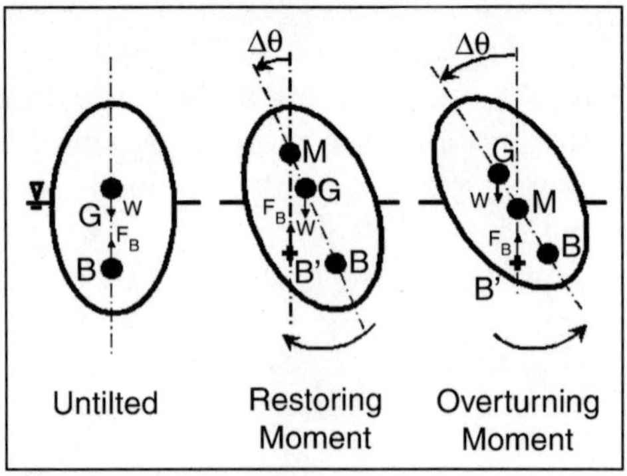

Consider a floating body tilted by an angle Dq, as shown below. For the untilted body the point G is the centre of gravity of the body where the body weight, W, acts. The point B is the centre of buoyancy (the centroid of the displaced volume of fluid) where the upward buoyancy force, F_B, acts.

When the body is tilted the centre of buoyancy moves to a new position, B', because the shape of the displaced volume changes. A new point, M, may be defined, called the Metacentre:

This is the point where a vertical line drawn upwards from the new centre of buoyancy, B', of the tilted body intersects the line of symmetry of the body. The buoyancy force, F_B, now acts through B'.

From the centre diagram in the figure we can see that W and F_B give a Restoring Moment that rotates the body back to its untilted position. From the right hand diagram in the figure we can see that W and F_B give an Overturning Moment that rotates the body even further in the tilted direction.

Hence, we can say; if the metacentre, M, lies above the centre of gravity, G, then the body is stable. In other words the Metacentric Height, MG, is positive ($MG = z_M - z_G > 0$). If the metacentre, M, lies below the centre of gravity, G, then the body is unstable. In other words the metacentric height, MG, is negative ($MG < 0$).

$$MG = MB - GB$$

$$MG = \frac{I}{V_s} - GB$$

The metacentric height, MG, is given by,

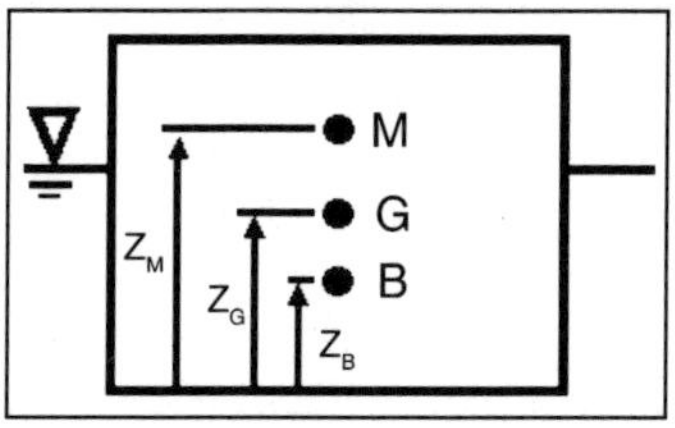

where I is the 2nd moment of area of the plan section of the body where it cuts the waterline. In other words, if you were to cut horizontally through the body at the water surface and look at the area of the body exposed by the cut, I is the 2nd moment of area of that body about the *longest* axis. V_S is the submerged volume (*i.e.* volume of fluid displaced) and GB is the distance between the centre of gravity and the centre of buoyancy ($= z_G - z_B$).

Strategy for solving buoyancy problems:

- From geometry of body and density of fluid and body equate; Weight of displaced fluid = Total weight of body. This gives the depth of immersion of the body or the weight of the body, whichever is unknown.
- To assess stability, first find the location of the centre of gravity G of the body.
- Then, find the location of the centre of buoyancy B (centroid of displaced volume). For a regularly shaped body this will be at half the height of the immersed portion of the body.
- Calculate the distance GB.
- Calculate MB, using $MB = I / V_S$. Note $I = p\ D^4/64$ for a circular section body and $bd^3/12$ for a rectangular section body (D is diameter, b and d are the sides of the rectangle).

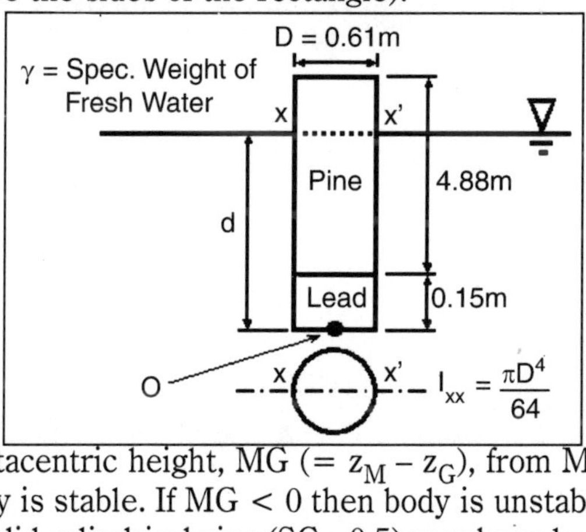

Calculate metacentric height, MG ($= z_M - z_G$), from MG = MB – GB. If MG > 0 then body is stable. If MG < 0 then body is unstable.

Example: A solid cylindrical pine (SG=0.5) spar buoy has a cylindrical lead (SG=11.3) weight attached as shown. Determine the equilibrium position (*i.e.*

depth of immersion) of the buoy in seawater (SG=1.03). Calculate the metacentric height and show that the buoy is stable.

Weight of fluid displaced = Weight of body d A 1.03 γ = 0.15 A 11.3 γ + 4.88 A 0.5 γ.

where A = π D^2 /4

Hence, d = 4.01 m,

$$MB = \frac{I}{V_S} = \frac{\frac{\pi D^4}{64}}{\frac{\pi D^2}{4} d} = \frac{D^2}{16\, d} = 5.8x10^{-3}\, m$$

Find centre of gravity of complete buoy by taking moments of the weight about the base (O):

$$GO = zG = \frac{0.15 \text{ A } 11.3\, \gamma\ (0.15/2) + 4.88 \text{ A } 0.5\, \gamma\ (4.88/2 + 0.15)}{0.15 \text{ A } 11.3\gamma + 4.88 \text{ A } 0.5\gamma}$$

$$= 1.56$$

Centre of buoyancy (BO) will be at half the height of the submerged part of the body (*i.e.* half the height of the displaced volume). Hence, BO=z_B=d/2 = 2.00m.

GB = GO – BO = 1.56 – 2.00 = -0.44 m

Therefore, metacentric height MG = MB – GB = 5.8 $\times$ 10^{-3} --0.44 = 0.45 m . MG > 0, hence, stable.

Now, think about how much longer the pine section of the buoy would have to be in order for it to become unstable? If the buoy is stable when MG > 0 and unstable when MG < 0, we can say that the limit between being stable or unstable is when MG = 0. In this case MB = GB. If you do the problem again, setting the length of the pine section as L (instead of 4.88m), and equating MB = GB, you can solve for L (by trial and error) to give the maximum length before instability sets in. Try it and see! [Answer = 7.20 m].

METACENTRIC HEIGHT

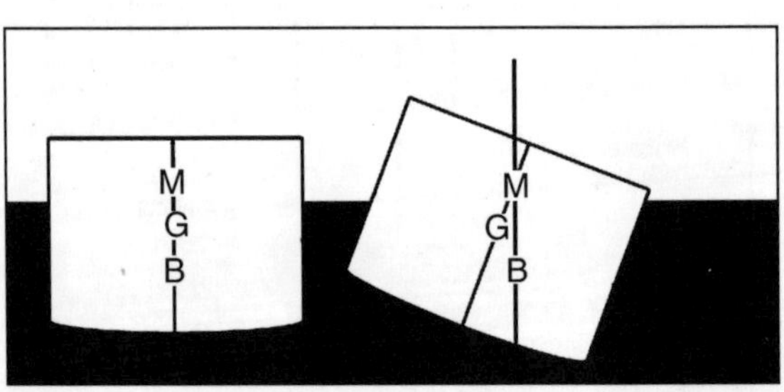

Ship Stability diagram, showing Center of Gravity (G), Center of Buoyancy (B), and Metacenter (M) with ship upright and heeled over to one side. Note that G is fixed, while B and M move as the ship heels.

The metacentric height (GM) is a characteristic of a ship which helps determine its stability in the water.

Different Centers

Initially the second moment of area increases as the surface area increases, increasing BM, so Mφ moves to the opposite side, thus increasing the stabilty arm. When the deck is flooded, the stability arm rapidly decreases.

The *center of buoyancy*, is the center of gravity of the volume of water which the hull displaces. This point is referred to as B in naval architecture. The center of gravity of the ship itself is known as G in naval architecture. When a ship is upright, the center of buoyancy is directly below the center of gravity of the ship.

The *metacenter* is the point where the lines intersect (at angle φ) of the upward force of buoyancy of φ ± dφ. When the ship is vertical it lies above the center of gravity and so moves in the opposite direction of heel as the ship rolls. The metacenter is known as M in naval architecture.

The distance between the center of gravity and the metacentre is called the metacentric height, and is usually between one and two metres. This distance is also abbreviated as GM. As the ship heels over, the centre of gravity generally remains fixed with respect to the ship because it just depends upon position of the ship's weight and cargo, but the surface area increases, increasing BMφ. The metacentre, Mφ, moves up and sideways in the opposite direction in which the ship has rolled and is no longer directly over the centre of gravity.

The righting force on the ship is then caused by gravity pulling down on the hull, effectively acting on its center of gravity, and the buoyancy pushing the hull upwards; effectively acting along the vertical line passing through the center of buoyancy and the metacenter above it. This creates a torque which rotates the hull upright again and is proportional to the horizontal distance between the center of gravity and the metacenter. The metacentric height is important because the righting force is proportional to the metacentric height times the sine of the angle of heel.

Righting Arm

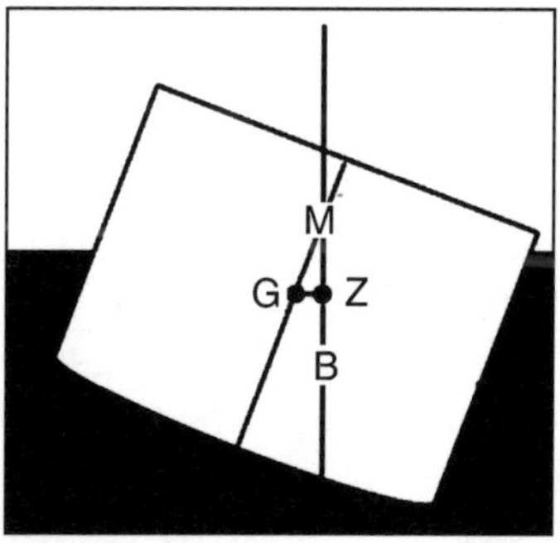

Distance GZ is the *righting arm*: a notional lever through which the force of buoyancy acts.

Sailing vessels are designed to operate with a higher degree of heel than motorized vessels and the righting torque at extreme angles is of high importance. This is expressed as the righting arm: the horizontal distance between the center of buoyancy and the center of gravity.

Monohulled sailing vessels are designed to have a positive righting arm (the *limit of positive stability*) at anything up to 120° of heel, although as little as 90° (masts flat to the surface) is acceptable. As the displacement of the hull at any particular degree of list is not proportional, calculations can be difficult and the concept was not introduced formally into naval architecture until about 1970.

Stability

GM and rolling period: GM has a direct relationship with a ship's rolling period. A ship with a small GM will be "tender" - have a long roll period - a low GM increases the risk of a ship capsizing in rough weather (see HMS Captain or HMS Vasa) and more likely to develop "synchronized rolling". It also puts the vessel at risk of potential for large angles of heel if the cargo or ballast shifts (see Cougar Ace).

If a ship with low GM is damaged and partially flooded the metacentric height will be reduced further and make it even less stable. On the other hand, a too large metacentric height can cause a vessel to be too "stiff"; excessive stability is uncomfortable for passengers and crew because it quickly snaps back upright after a wave or wind gust which heeled it over has passed. An overly stiff vessel rolls with a short period and high amplitude.

This can lead to damage to the ship and cause cargo to break loose or shift. A passenger ship will typically have a long rolling period for comfort, perhaps 28 seconds while a tanker or freighter might have a rolling period of 13 to 15 seconds.

Damaged Stability

If a ship floods the loss of stability is due to the free surface effect, as the water accumulating in the hull will be in the bilges, lowering the centre of gravity and actually increasing the metacentric height. This additional mass will however reduce freeboard (Distance from water to the deck) and the ships angle of down flooding (minimum angle of heel at which water will be able to flow into the Hull).

The range of positive stability will be reduced to the angle of down flooding resulting in a reduced righting lever.

When the vessel is inclined the fluid in the bilge will move to the lower side, shifting its centre of gravity towards the list, further extending the heeling force. This is known as the free surface effect.

FREE SURFACE EFFECT

In tanks or spaces that are partially filled with a fluid or semi-fluid (fish, ice or grain for example) as the tank is inclined the surface of the liquid, or semi-fluid, stays level. This results in a displacement of the centre of gravity of the tank or space. The effect is similar to that of carrying a large flat tray of water. When an edge is tipped, the water rushes to that side which exacerbates the tip even further.

The significance of this effect is proportional to the square of the width of the tank or compartment, so two baffles separating the area into thirds will reduce the displacement of the centre of gravity of the fluid by a factor of 9. This is always of significance in ship fuel tanks or ballast tanks, tanker cargo tanks, and in flooded or partially flooded compartments of damaged ships. Another worrying feature of free surface effect is that a positive feedback loop can be established, in which the period of the roll is equal or almost equal to the period of the motion of the centre of gravity in the fluid, resulting in each roll increasing in magnitude until the loop is broken or the ship capsizes.

TRANSVERSE AND LONGITUDINAL METACENTRIC HEIGHTS

There is also a similar consideration in the movement of the metacentre forward and aft as a ship pitches. Metacentres are usually separately calculated for transverse (side to side) rolling motion and for lengthwise longitudinal pitching motion. These are variously known as $\overline{GM_T}$ and $\overline{GM_L}$, GM(t) and GM(l), or sometimes GMt and GMl .

Technically, there are different metacentric heights for any combination of pitch and roll motion, depending on the moment of inertia of the waterplane area of the ship around the axis of rotation under consideration, but they are normally only calculated and stated as specific values for the limiting pure pitch and roll motions.

Measuring Metacentric Height

The metacentric height is normally estimated during the design of a ship but can be determined by an inclining experiment or Inclining test once it has been built. This can also be done when a ship or offshore floating platform is in service. It can be calculted by theoretical formulas based on the shape of the structure.

STABILIZER SOLUTIONS

Because ships under way have their own characteristics and natural rolling periods, the amplitude can to a certain extent be reduced with the fitting of bilge keels, anti-rolling tanks, fin stabilizers, or a combination of these. Forces are generated, with the help of computers, equal and opposite to those of the sea. Most passenger ships make use of the equipment.

STABILITY OF IMMERSED AND FLOATING BODIES

STABILITY OF FLOATING BODIES

A floating body is STABLE if, when it is displaced, it returns to equilibrium. A floating body is Unstable if, when it is displaced, it moves to a new equilibrium.

Consider a floating body tilted by an angle Dq, as shown below. For the untilted body the point G is the centre of gravity of the body where the body weight, W, acts. The point B is the centre of buoyancy (the centroid of the displaced volume of fluid) where the upward buoyancy force, F_B, acts.

When the body is tilted the centre of buoyancy moves to a new position, B', because the shape of the displaced volume changes. A new point, M, may be defined, called the Metacentre. This is the point where a vertical line drawn upwards from the new centre of buoyancy, B', of the tilted body intersects the line of symmetry of the body. The buoyancy force, F_B, now acts through B'.

From the centre diagram in the figure we can see that W and F_B give a Restoring Moment that rotates the body back to its untilted position. From the right hand diagram in the figure we can see that W and F_B give an Overturning Moment that rotates the body even further in the tilted direction.

Hence, we can say; if the metacentre, M, lies above the centre of gravity, G, then the body is stable. In other words the Metacentric Height, MG, is positive ($MG = z_M - z_G > 0$). If the metacentre, M, lies below the centre of gravity, G, then the body is unstable. In other words the metacentric height, MG, is negative ($MG < 0$).

The metacentric height, MG, is given by:

$$MG = MB - GB \qquad \text{or} \qquad MG = \frac{I}{Vs} - GB$$

where I is the 2nd moment of area of the plan section of the body where it cuts the waterline (this is the solid plane surface you'd see if you cut horizontally through a solid body at the water surface lifted the top part up and looked at the bottom of it!), V_S is the submerged volume (*i.e.* volume of fluid displaced) and GB is the distance between the centre of gravity and the centre of buoyancy ($= z_G - z_B$).

Strategy for solving buoyancy problems:

- From geometry of body and density of fluid and body equate; Weight of displaced fluid = Total weight of body. This gives the depth of immersion of the body or the weight of the body, whichever is unknown.
- To assess stability, first find the location of the centre of gravity G of the body.

- Then, find the location of the centre of buoyancy B (centroid of displaced volume). For a regularly shaped body this will be at half the height of the immersed portion of the body.
- Calculate the distance GB.
- Calculate MB, using MB = I / V_S. Note I = p $D^4/64$ for a circular section body and $bd^3/12$ for a rectangular section body (D is diameter, b and d are the sides of the rectangle).
- Calculate metacentric height, MG (= $z_M - z_G$), from MG = MB – GB. If MG > 0 then body is stable. If MG < 0 then body is unstable.

Example: A solid cylindrical pine (SG=0.5) spar buoy has a cylindrical lead (SG=11.3) weight attached as shown. Determine the equilibrium position (*i.e.* depth of immersion) of the buoy in seawater (SG=1.03). Calculate the metacentric height and show that the buoy is stable.

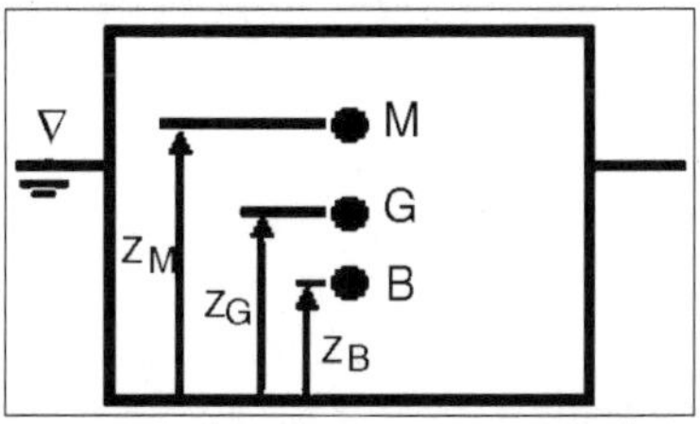

$$\text{Weight of fluid displaced} = \text{Weight of body}$$

$$d\,A\,1.03\,g = 0.15\,A\,11.3\,g + 4.88\,A\,0.5\,g$$

$$\text{where } A = p\,D^2/4$$

$$\text{Hence, } d = 4.01\text{ m}$$

Find centre of gravity of complete buoy by taking moments of the weight about the base (O):

$$GO = z_G = \frac{0.15A11.3\gamma(0.15/2)+4.88A0.5\gamma(4.88/2+0.15)}{0.15A11.3\gamma+4.88A0.5\gamma} = 1.56\text{m}$$

$$0.15\,A\,11.3\,g + 4.88\,A\,0.5\,g$$

Centre of buoyancy (BO) will be at half the height of the submerged part of the body (*i.e.* half the height of the displaced volume). Hence, BO=z_B=d/2 = 2.00m.

$$GB = GO - BO = 1.56 - 2.00 = -0.44\text{ m}$$

Therefore, metacentric height MG = MB – GB = $5.8\text{x}10^{-3}$ - -0.44 = 0.45 m . MG > 0, hence, stable.

Now, think about how much longer the pine section of the buoy would have to be in order for it to become unstable? If the buoy is stable when MG > 0 and unstable when MG < 0, we can say that the limit between being stable or unstable is when MG = 0. In this case MB = GB. If you do the problem again, setting the length of the pine section as L (instead of 4.88m), and equating MB = GB, you can solve for L (by trial and error) to give the maximum length before instability sets in. Try it and see! [Answer = 7.20 m].

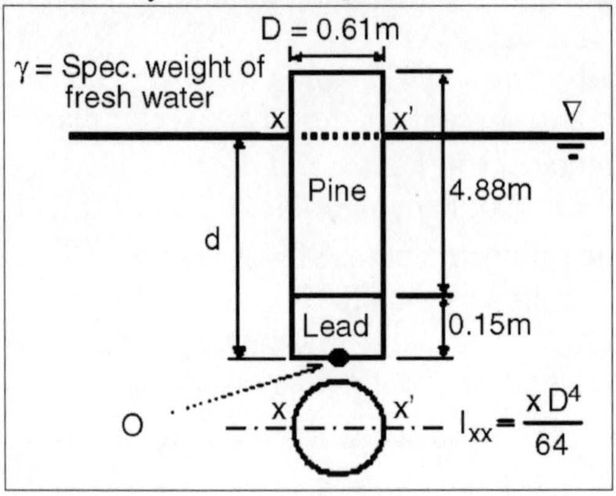

We have expanded and edited this page to provide both more theory and Worked Examples. We hope that you will find the changes an improvement.

We are always happy to receive any comments that you may have on layout or content. The new pages will be found under "Engineering/ Fluid Mechanics/ Ships and other Floating Bodies"

CENTRE OF BUOYANCY AND STABILITY

The Buoyancy Force act through the Centre of Gravity of the Displaced Fluid and is called " The Centre of Buoyancy"

There are three Types of Equilibrium:

- *Stable*: The body returns to it's original position if given a small angular displacement.
- *Neutral*: The body remains in a new position if given a small angular displacement.
- *Unstable*: The body heals further over if given a small angular displacement.

STABILITY OF FULLY SUBMERGED BODIES

$$MB = \frac{I}{V_s} = \frac{\frac{\pi D^4}{64}}{\frac{\pi D^2}{4}} = \frac{D^2}{16d} = 5.8 \times 10^{-3} \text{m}$$

Let

- V = Volume of Body
- w = Specific weight of the fluid.
- W = Weight of the Body

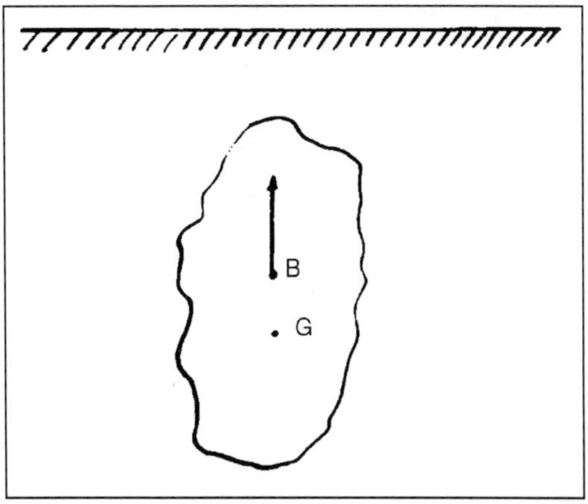

- G is the Centre of Gravity and B is the centre of Buoyancy

If B and G are coincident the the Body will be in Neutral Equilibrium.

- B is below G then the Body is in Unstable equilibrium.
- B is above G then the body is in Stable equilibrium.

This last case has a righting couple of W B GWhere θ is the angle of tilt. Note that the Position of B and G relative to the Body does not change when the body is rotated.

STABILITY OF PARTIALLY SUBMERGED BODIES

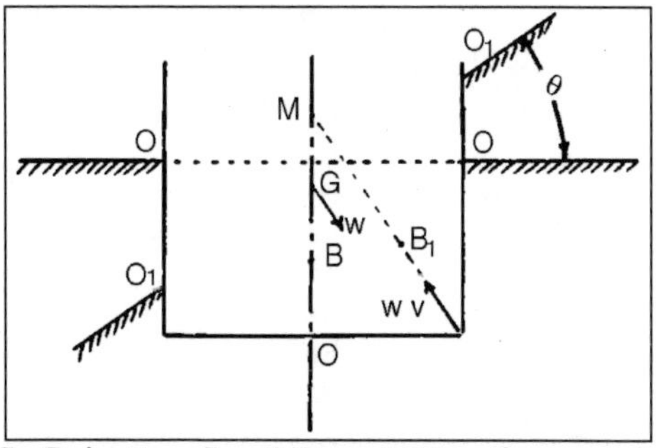

G is the C of G. B is the C. of B and The line o o is the original water surface. After tilting O' O' is the new water line and the angle of Tilt is θ G remains in the same position relative to the ship but the Centre of Buoyancy moves to B_1

M is the "Meta Centre" and is defined as the point where the vertical through the new Centre of Buoyancy meets the original vertical through the Centre of Gravity after a very small angle of rotation.

MG is called the metacentric height:

Therefore for stable equilibrium for a floating Partially Submerged Body the Metacentre must be above the Centre of Gravity G. If the Metacentric height is zero the Body will be in Neutral equilibrium. The Righting couple W GM sinθ

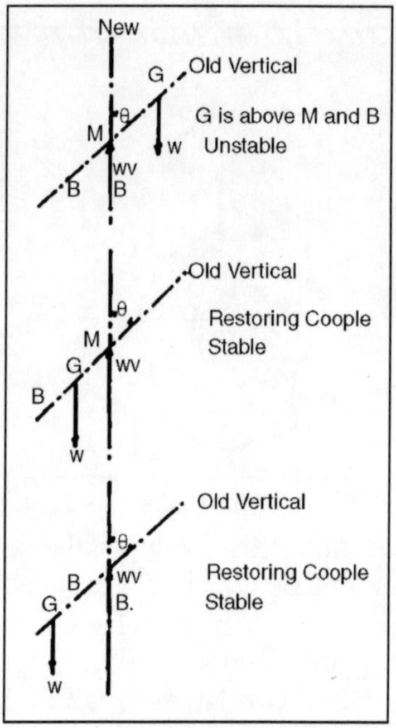

DETERMINATION OF METACENTRIC HEIGHT

Experimentally

Let W be the weight of the Boat plus it's Load:

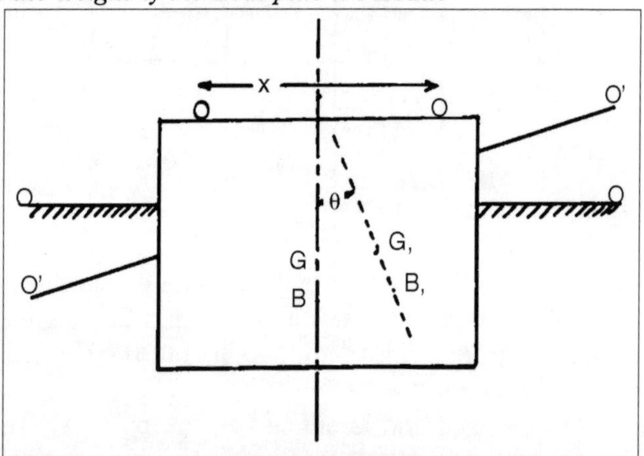

A small load w is moved a distance x and causes a tilt of angle θ. The Boat is now in a new position of equilibrium with B' and G' lying along the Vertical through M.

The Moment due to the movement of the load is given by:

wx = W × GM θ = moment due to movement of C of G

$$\therefore \text{GM} = \frac{\text{wx}}{\text{W}\theta}$$

Theory,

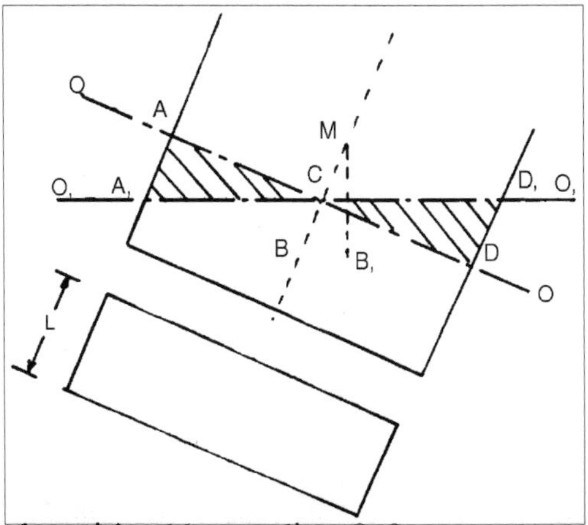

The Ship tilts from it's old waterline O O to a new waterline O' O' as it moves through an angle θ. Due to the movement of the wedge of water from A' A C to D' D C, the Centre of Buoyancy moves from B to B'

The Change in the moment of the buoyancy Force = wV X B B'

= w V × BM θ where θ is small

The Volume of the Wedge A C A'

$$=\frac{1}{2}\times\frac{b}{2}\times\frac{b}{2}q\times L=\frac{b^2L}{8}\theta$$

Therefore the Moment of the Couple due to the movement of the wedge,

$$w\times\frac{wb^2L\theta}{8}\times\frac{2}{3}b$$

$$wV\times BM\theta=\frac{wb^3L}{8}\theta$$

$$BM=\frac{b^3L}{12V}=\frac{I}{V}$$

Where,

I is the Second Moment of Area of the Water Plane Section

V is the volume of water Displaced.

Thus if the positions of B and G are known or can be calculated, then the distance GM can be determined since: 0

GM = BM + BG = Bm + BO – OG

There are in fact two Metacentric heights of a ship. One for Rolling and the other for Pitching. The former will always be less than the latter and unless otherwise stated, the Metacentric given will be for Rolling.

A Pontoon measuring 20ft.by 12ft. and 4ft deep, weighs 12 tons. It carries a load of 8 tons. The Pontoon sits in sea water with a density of 64 lb/cu. ft.

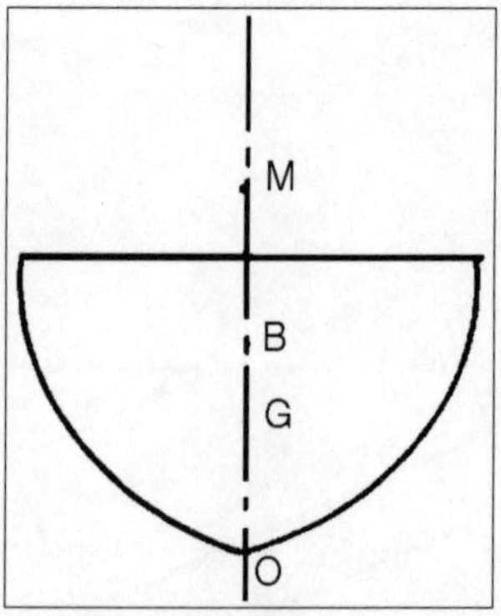

Metacentric Height For A Vessel With Liquid Ballast.

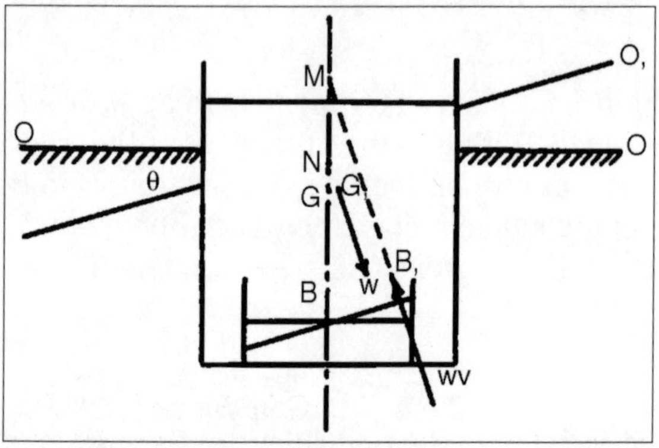

- O O is the original Waterline
- G is the Centre of Gravity of the Ship and Ballast.
- B is the Centre of Buoyancy.

The Ship is tilted through a small angle θ Clockwise giving a new waterline of O' O'

Due to the movement of the water wedge the Centre of Buoyancy of the Ship (*i.e.* The C of G of the displaced liquid) moves to B'.

By the Previous,

$$\text{Theory BM} = \frac{I}{V}$$

Where,

I is the second Moment of Area of the Water plane area

V is the Total displacement by the Ship and it's contents.

Due to the movement of the liquid in the Ballast tank the C of G of the Ship and Ballst moves to G'. Now the vertical through G' cuts the old vertical centre line at N. The Stability of the ship depends upon whether N is above or below M.

The Metacentric Height is now MN,

Consider the Liquid in the Ballast Tank.

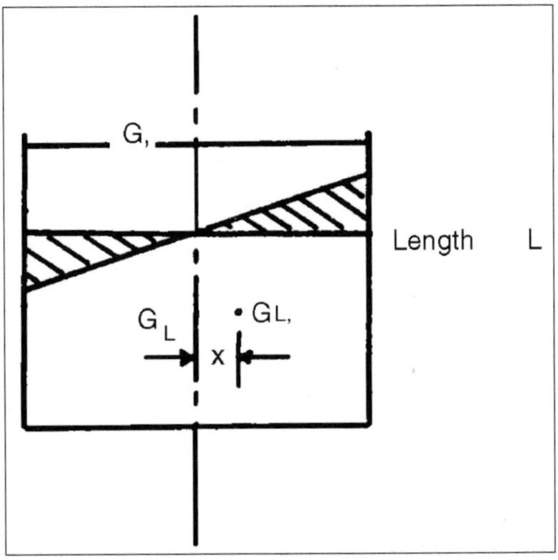

Let,

- I' = The second moment of area of the liquid surface about it's centre line.

$$= \frac{L'B'^3}{12}$$

- V' = the Volume of liquid in the Tank,
- w' = the specific weight of the liquid.
- x = The movement of the C of G in the tank due to the tilting of the vessel.

The moment of the Couple due to the movement of the Wedge = The Moment of the Couple due to the movement of the C of G.

$$w'v'\ x = w' \times \frac{1}{2} \times \frac{b'}{2} \times \frac{b'\theta}{2} \times L' \times \frac{2b'}{3}$$

$$\therefore \quad V'x = \frac{L'b'^3}{12}$$

$$\therefore \quad x = \frac{I'\theta}{V'}$$

Thus the C of G of the ship and Contents moves from G to G' where G G' is given by:

$$W \times GG' = w'V' \times x$$

or

$$wv \times GG' = w'V' \times x$$

$$\therefore GG' = \frac{w'V'}{wV} \times \frac{I}{V'} \times \theta = \frac{w'I\theta}{wV}$$

But GG' = GN × θ

$$\therefore GN = \frac{w'I}{wV}$$

Therefore the Metacentric Height MN = MB + BO - OG - GN:

$$= \frac{I}{V} + BO - OG - \frac{w'I'}{wV}$$

Thus the Metacentric Height has been reduced by the liquid ballast by an amount,

$$\frac{w'I'}{wV}$$

Divided Ballast Tanks

If the Ballast Tank is divided into two by a longitudinal Partition, the the above equation becomes.

$$x = \frac{w_1 I_1}{wV} \text{ and } y = \frac{w_2 I_2}{wV}$$

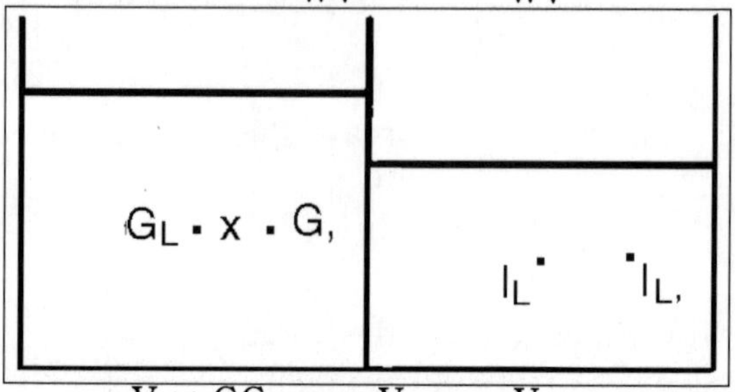

$$wV \times GG_1 = w_1 V_1 \times w_1 V_2 \times y$$

$$\therefore \quad GG_1 = \frac{w_1 \theta}{wV}(I_1 + I_2)$$

$$\therefore \quad GN = \frac{w_1}{wV}(I_1 + I_2)$$

Comparing the two separate cases:

Single Tank of width b.

$$I_1 = \frac{L_1 b_1^3}{12}$$

Same tank divided into two equal compartments.

$$I_1 + I_2 = \frac{2 \times L_1}{12} \times \left(\frac{b_1^3}{2}\right)^3 = \frac{1}{4} \times \frac{L_1 b_1^3}{12}$$

This shows that the reduction in the Metacentric height that occurs in case two is a quarter that experienced in case one. For this reason the tanks in Oil Tankers are divided into small compartments.

Right Couple With A Liquid Ballast.

The Righting Couple due to the Buoyancy Force acting upwards through B_1 and M and the weight of the ship acting downwards through G_1 and N

$= W \times MN \times \theta$ where θ is small

This is the same as in the simple case without a liquid balance but it must be remembered that the Metacentric Height is now MN and NOT MG

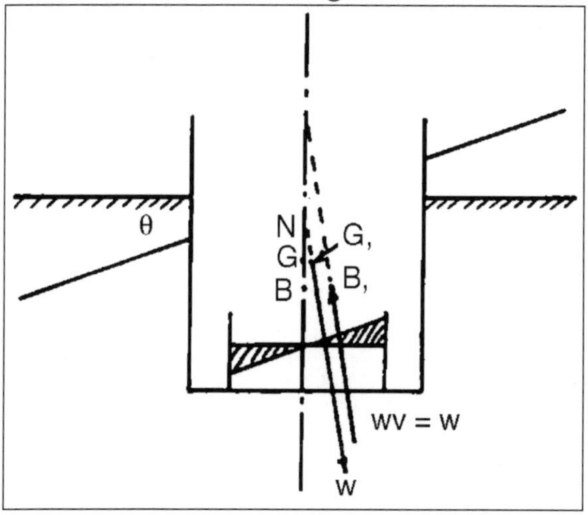

Example:

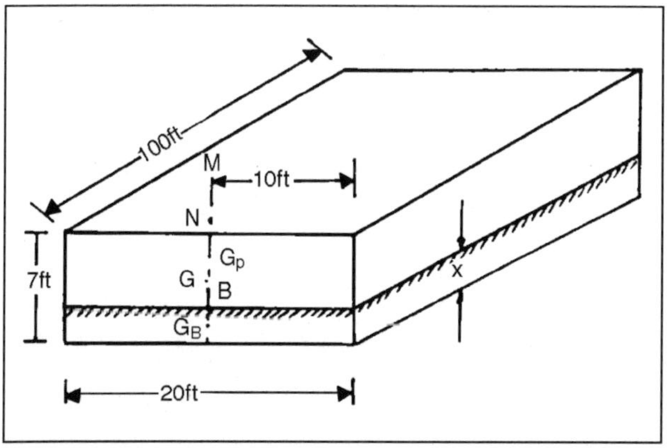

Period of Oscillation of a Floating Body

- Torque = Moment of Inertia × Angular Acceleration.
- The Solution of,

$$\frac{d^2y}{dx^2}+a^2y=0$$

is given by = A sin ax + B cos ax

For small angles of heel, the Body can be regarded as Oscillating about it's Metacentre, in a manner similar to a Pendulum about it's point of suspension.

If,

- W = The weight of the Ship
- M = The Metacentric Height

- k = The Radius of Gyration of the Ship about a horizontal Axis through the CofG.
- θ = The Rotation after a time t.

Angular acceleration,

$$= \frac{d^2\theta}{dt^2}$$

(Direction towards equilibrium positioin)

The Moment of Inertia of the ship about the C of G is,

$$\frac{w}{g}k^2.$$

The Moment of Inertia of the ship about M is,

$$\frac{W}{g}(k^2 + m^2)$$

Provided that m is small compared with k, these two can be regarded as equal,

$$L_G \approx I_M$$

For small angles of heel, the Righting Couple = W m θ

$$\therefore \text{W m q} = \text{Im} = \frac{d^2\theta}{dt^2} \approx I_G \times \frac{d^2\theta}{dt^2}$$

$$= -\frac{W}{g}k^2 \frac{d^2\theta}{dt^2}$$

$$\therefore \frac{d^2\theta}{dt^2} + \frac{gm}{k^2\theta} = 0$$

The Solution of this Equation is given by:

$$\theta = A \sin\sqrt{\frac{gm}{k^2}}t + b\cos\sqrt{\frac{gm}{k^2}}t$$

When t = 0 θ = 0 and therefore B = 0

When θ = o and t = T/2 *i.e.* T = the time of a complete oscillation.

$$A \sin\sqrt{\frac{gm}{k^2}} \times \frac{T}{2} = 0$$

Since $A \neq 0 \sin\sqrt{\frac{gm}{k^2}} \times \frac{T}{2} = 0$

The simplest solution:

$$\sqrt{\frac{gm}{k^2}} \times \frac{T}{2} = \pi$$

$\therefore$ The Periodic Time $T = 2\pi\sqrt{\frac{k^2}{mg}}$

Bibliography

A.K. Shrivastava : *Encyclopaedia of Environmental Pollution, Agriculture and Health Hazards (5 Vols-Set)*, APH Publication, Delhi, 2004.

Ajit K. Ghosh.: *A Modern Book on Plant Biotechnology*, Cyber Tech Publication, New Delhi, 2009.

Amitava Rakshit and Priyankar Raha.: *Terminologies in Agricultural Sciences*, IBDC Publication, Delhi, 2009.

Ashok Kumar Sharma : *A Textbook of Biochemistry*, Random Publication, Delhi, 2012.

B Balaswamy : *Communication for Sustainable Development*, Concept Publisher, Delhi, 2006.

Bidhan Roy and Asit Kumar Basu.: *Abiotic Stress Tolerance in Crop Plants : Breeding and Biotechnology*, New India Publication, Delhi, 2009.

Brajendra, Kailash Chand Verma, Gyan Prakash Mishra, P. Muthuraman.: *Objective Agricultural Science*, New India Publication, Delhi, 2009.

Bruno Dorin and Thomas Jullien : *Agricultural Incentives in India* : Challenges, Oxford University Press, Delhi, 2005.

C.P. Malik, Bhavneet Kaur and Chitra Wadhwani.: *Advanced Topics in Biotechnology and Plant Biology*, MD Publication, Delhi, 2008.

C.R. Sudharmai Devi : *Analytical Procedures in Soil Science and Agricultural Chemistry*, Agrotech Publisher, Delhi, 2004.

Dewasish Choudhary : *Basics of Agricultural Chemistry*, Anmol Publication, Delhi, 2009.

G Byju : *Soil Chemistry : Basic Concepts*, New India Publication, Delhi, 2008.

Harry H. Love : *Application of Statistical Methods to Agricultural Research*, Cosmo Publication, Delhi, 2006.

J.B. Ambekar : *Agricultural Communication and Sustainable Development*, Shruti Publication, Delhi, 2010.

Jennifer A. Elliott : *An Introduction to Sustainable Development*, Routledge Publsher, Delhi, 2010.

K.S. Dadhwal and Mohit Gahlot : *Advances in Agricultural Research in India*, International Book Distribut, Delhil, 2002.

N.K. Prasad : *Soil Fertility and Plant Nutrition, IBDC Publishers*, Delhi, 2013.

P N Mishra : *Environment Pollution and Its Control*, Sumit Enterprises, Delhi, 2007.

R.A.S. Tomar : *Encyclopaedia of Agricultural Chemistry*, Anmol Publication, Delhi, 2010.

Shivanand Tolanur : *Practical Soil Science and Agricultural Chemistry*, International Book Distributing, Delhi, 2004.

Sriniwas Puri : *Soil Chemistry*, Oxford Book, Delhi, 2010.

T.V.S. Prasad : *Soil Chemistry : Nutrient and Water Management in Agricultural Soils*, Dominant Publication, Delhi, 2009.

Tapan Adhikari, S. Kundu and A. Subba Rao : *Nanotechnology in Soil Science and Plant Nutrition*, New India Publishing Agency, Delhi, 2013.

Index